D0450120

THE LOOP

THE LOOP

HOW TECHNOLOGY IS CREATING A WORLD WITHOUT CHOICES AND HOW TO FIGHT BACK

JACOB WARD

NEW YORK

Cover design by Amanda Kain
Cover image © jani_arh/Shutterstock
Cover copyright © 2022 by Hachette Book Group, Inc.

Portions of this book were originally published, in different form, on Medium's "NewCo."

Hachette Books
Hachette Book Group
1290 Avenue of the Americas
New York, NY 10104
HachetteBooks.com
Twitter.com/HachetteBooks
Instagram.com/HachetteBooks

First Edition: January 2022

Published by Hachette Books, an imprint of Perseus Books, LLC, a subsidiary of Hachette Book Group, Inc. The Hachette Books name and logo is a trademark of the Hachette Book Group.

The Hachette Speakers Bureau provides a wide range of authors for speaking events. To find out more, go to www.hachettespeakersbureau.com or call (866) 376-6591.

The publisher is not responsible for websites (or their content) that are not owned by the publisher.

Print book interior design by Trish Wilkinson

Library of Congress Cataloging-in-Publication Data
Names: Ward, Jake, author.
Title: The loop : how technology is creating a world without choices and how to fight back / Jacob Ward.
Description: First edition. | New York : Hachette Books, 2022. | Includes bibliographical references.
Identifiers: LCCN 2021034378 | ISBN 9780316487207 (hardcover) | ISBN 9780316487221 (ebook)
Subjects: LCSH: Artificial intelligence. | Brain. | Decision making.
Classification: LCC Q335 .W358 2022 | DDC 006.3—dc23
LC record available at https://lccn.loc.gov/2021034378

ISBNs: 9780316487207 (hardcover), 9780316487221 (ebook)

Printed in the United States of America

LSC-C

Printing 1, 2021

For Julie,
who saved me

For Josephine and Juniper,
who will save us all

On their way toward modern science human beings have discarded meaning. The concept is replaced by the formula, the cause by rules and probability.

—Max Horkheimer and Theodor W. Adorno,
Dialectic of Enlightenment, 1947 (trans. Edmund Jephcott)

CONTENTS

THE LOOP

Introduction

THE GENERATION SHIP

WHENEVER OUR FUTURE on this planet looks bleak, we can't help but think about other planets. We've spilled over our place in nature, and we can't seem to get along well enough to agree on our shared salvation. Let's go somewhere else and start over. How far away is the next habitable planet, anyway?

Interplanetary scientists despise the notion of translating light-years to a speed that you and I can grok as a number, the way we put a number to the speed of our own cars, so it always falls to journalists like me to thumbnail a best guess as to just how long it might take to carry us that same distance. Bear with me, please.

Our current rocket technology can propel a ship through space at speeds of roughly 20,000 miles per hour. By the standards you and I are used to, that's incredibly fast. In our own atmosphere the friction against the surrounding air at that speed would melt through any material we've invented and would incinerate you and me before we could even strike up a conversation about where we were going and what we wanted to build when we got there. But let's use that speed as a benchmark, because in space that speed is a terribly slow rate to cross the enormous distances between planets.

Mars, whose orbit is directly adjacent to our own, is the most survivable other planet in our solar system. But that's not saying much. Sure, other planets are more horrible. Piercing the thirty-mile layer of clouds that surrounds Jupiter, our rocket's engines would begin to either choke or over-fire in its flammable nightmare of hydrogen and helium, and then would die completely as they hit the liquid form of the same stuff roughly 13,000 miles past the cloud cover that keeps us from knowing anything about what's beneath that poisonous ocean. That crew would drown (or maybe drown *and* burn, unheard of on Earth) without ever making it out of the ship.

Mars is comparatively pleasant. For one thing, there's stable footing. And a nice day on Mars might actually *feel* nice. With the sun high in the sky, you'd enjoy temperatures as high as 68° F, a clear August afternoon in San Francisco or Johannesburg. But if you happened to exit the ship at one of the poles, at night, the temperature could be less than –200° F, cold enough to not only instantly kill you, but also almost instantly freeze your limbs into brittle branches that would shatter under a hammer blow. And let's not forget that even in the balmiest regions of the planet, there's nothing to breathe, so you're not getting far on even the most pleasant day. You'd bound perhaps fifty yards in the low gravity before you could no longer hold your breath, then hypoxia would disorient you, and you couldn't make it back to the oxygen of the ship. You'd wind up unconscious and twitching, your heart would stop, and the red dust would slowly settle around your corpse.

That's why scientists and journalists alike are so excited about exoplanets, the term for planets beyond our solar system that seem to offer the possibility of a livable atmosphere and surface. Humanity has been treated in the last few years to a steady stream of optimistic fantasy destinations emanating from the now-defunct Kepler space telescope. Kepler essentially squinted out into deep space to see how light from distant stars bent around intervening planets too far away to image in any detail. Depending on how the light goes around a planet on its way to us, astrophysicists can calculate not only the size of that planet, but how far it is from the source of light, meaning we can determine whether the relation between the planet's size and the distance to its star possibly indicates that planet might host some sort of atmosphere.

The Kepler mission measured the light from roughly 150,000 stars and found several hundred planets whose ratio of size and star-distance makes them candidates—just candidates, but a real possibility—for human respiration and occupancy. Land, walk around, build a cabin out of whatever materials exist, just imagine! And when we consider the vast distances of space, the closest exoplanets are, in fact, relatively close.

But before we pop the champagne and pour our savings into SpaceX, let's think about what it takes to get to another planet. A trip to Mars, for instance, is comparatively brief. Depending on where it and Earth are in their orbits, the journey could be made in between three hundred and four hundred days. But humans have never traversed open space for that amount of time. The journey to the moon currently takes about seventy-two hours, and astrophysicists and medical experts quietly point out in private conversation that it's a miracle none of the two dozen people who went to the moon died during the trip. A trip to Mars would involve exposing the crew to the dangers of deep space for roughly a full year. And those dangers go on and on. Deadly amounts of radiation permeate everything in the deep blackness between planets. Space is full of dirt and grit that could disable the ship. (A whole field of astrophysics studies the interstellar medium and has shown that if you held a gloved white hand out the window of the ship as one does on the highway, it would come back blackened by, well, whatever it is that's out there.) Also consider that if a mishap killed the ship and the crew, the event would be torturously broadcast, on time delay, to the whole of Earth, presumably killing off our species' desire to travel to Mars in the process.

And even if all goes well for the crew, that's a long time confined together in a space no bigger than a vacation rental, as all of us who spent the pandemic year locked in with family, roommates, or alone know too well. In fact, before the coronavirus made test cases of us all, psychologists and logisticians who worried about a Martian crew driving each other nuts spent time observing actual astronauts confined in these sorts of tiny spaces for the duration, either of a simulated trip to Mars or a stay on the planet. And it hasn't gone well. On almost every sardine-style simulation someone has suffered serious injury or illness. A simulated Martian habitat on a military base on the island of Hawaii has seen a half-dozen such missions over the years, including one where a crew member had to withdraw for medical reasons (the mission's organizers haven't publicly revealed what it was). Seeking to learn from the experience, the crew pretended to treat their missing member as dead and enacted setting a fake body out on the simulated Martian tundra, where it would be perfectly preserved for a journey back to Earth for burial. In the final Hawaiian mission, the simulation was compromised when one of the crew was electrified by live wiring, and earthly paramedics had to come inside and drive the crew member away in an ambulance. But putting aside the physical danger of living isolated on Mars, the missions have revealed that . . . people get weird. "You can select a crew all you want, get the right fit and mix, but there's too many variables when it comes to human beings," a psychologist for the mission told the *Atlantic.* "It's just really hard to predict how we're going to perform in all situations."

That's just Mars. It's next door to us, cosmically speaking. Now imagine how weird we'd become trying to reach the nearest *exoplanet.*

Let's imagine we're standing together on the launch pad at NASA's Cape Canaveral facility near Orlando, and staring up at the stars together. As I write this, the last constellation above the horizon is Centaurus. The centaur's front hoof is a bright star. In fact, it's three stars—a pair called Alpha Centauri A and B, and, dimmest of the trio, Proxima Centauri. Here, look through this telescope. See? You can tell them apart. But what we can't see is that there is, in fact, a planet circling the faint light of Proxima Centauri. Man, I *wish* we could see it. Because that planet, Proxima Centauri b, is the nearest known exoplanet to Earth.

We have no idea what life would be like on Proxima Centauri b, or what the place even looks like. There may be many reasons that it just won't work for human habitation. It's possible that stellar winds may push so much radiation across its surface that we'd all be poisoned before we got the first shelter built, and those same winds may have stripped away any breathable atmosphere, meaning we'd have to live underground. It's also possible that the planet's orbit of Proxima Centauri happens at such a cadence that one side of the planet permanently faces the sun, meaning half of the planet is always daylit, and the other is always in darkness.

But let's stay hopeful. Let's imagine that it's a perfectly habitable place, with warm winds and a liquid ocean and strange, vivid landscapes of rock and vegetation and alien snow. Let's go there!

First, the good news. Proxima Centauri b is only 4.2 light-years away. That means that light, the fastest thing we know of, at roughly 186,000 miles per second, would take only 4.2 years to streak from our planet to Proxima Centauri b's weird, wild shores. For photons, that's a very short trip.

The bad news is that for humans, it's a very long trip. We don't travel at that speed. Not even close. We'll need much more time to get there. In fact, it's so much time that no one who has ever set foot on Earth will ever set foot on Proxima Centauri b.

If we were to board a spacecraft and ride it from the outer edge of our atmosphere all the way to Proxima Centauri b, you and I, who boarded the ship fit and trim, chosen as we were from billions of applicants, would die before the voyage reached even 1/100th of the intervening distance. It's such an outrageously long journey that a human life span is just a tiny fraction of the time it will take.

Here's the napkin math. At a speed of 20,000 miles per hour—the speed of our top-performing modern rockets—4.2 light-years translates to more than 130,000 *years* of space travel.

One hundred thirty *thousand* years. This means that the time involved to reach our closest exoplanet neighbor would crush us, and our children, and their children into dust a thousand times over before anyone had a chance to breathe alien air.

Could we put ourselves in some sort of coma for the journey, as the characters in *2001*, *Alien*, and *Interstellar* do? Trauma surgeons, experimenting with the same concept that inspired Ted Williams's family to freeze his head for possible transplant in the future, are currently experimenting with procedures that can revive a semi-frozen patient after two hours without a pulse. But we're a long way from freezing ourselves for as long as this might take. Using current technology, anywhere from 900 to 1,300 human generations would pass on the way to Proxima Centauri b before the ship arrives. *Generations*. So how will we ever get there? A generation ship.

First proposed in varying forms by early rocket pioneers, science fiction writers, and astrophysicists with a few beers in them, the general notion is this: get enough human beings onto a ship, with adequate genetic diversity among us, that we and our fellow passengers cohabitate as a village, reproducing and raising families who go on to mourn you and me and raise new children of their own, until, thousands of years after our ship leaves Earth's gravity, the distant descendants of the crew that left Earth finally break through the atmosphere of our new home.

I once had dinner with an evolutionary biologist, and I asked him what it is about Darwin's theory that the human mind has most trouble seeing clearly. "It's the outrageously long periods of time involved," he said without hesitation. "We're just not equipped to be able to imagine the amount of time evolution takes."

That inability to see evolution's progress on the vast plain of time also means that planning for ongoing communal life aboard a single spaceship is largely beyond our natural gifts. I pat myself on the back when I get out ahead of birthday presents for my children. The ability to plan an environment that will keep my great-grandchildren alive and happy enough to reproduce with the great-grandchildren of my colleagues is an entirely separate matter.

Think of the logistics: How will we create new clothes a thousand years after the first boxes of cloth have collapsed and moldered away? Think of nutrition: What crop lasts even 100 years, much less 100,000 years? And relationships: Will assortative mating, the natural tendency to band together with people that most resemble us, continue onboard, cutting into the genetic diversity of offspring and wiping out that bloodline after a few succeeding generations? (Speaking of factions, I have half a screenplay written about the tribal nightmare that such a ship might become. Movie producers: I'm easy to find online. Hit me up. You only need to build one set!)

Consider this: humans only spread across the planet roughly 140,000 years ago. That's when our species stood up, looked around, and accessed some higher form of ambition and curiosity in deciding that it was worth risking everything to leave what is now the African continent and look for game and water elsewhere. Every human accomplishment you've ever read about—language, religion, the Egyptian, Greek, and Roman empires—has happened since then. We've only been "modern humans" for that long. And that means that a generation ship isn't just about finding and training the right one hundred people to seed a thousand-generation journey. It's about literally bottling the equivalent of our entire history as a species for a repeat performance in captivity.

A generation ship is every sociological and psychological challenge of modern life squashed into a microcosmic tube of survival and amplified—generation after generation. Will the final generation be able to function well enough to even set the ancient ship, rebuilt in flight over and over, down on the surface of humanity's new home? How do we possibly plan for this sort of time scale?

The idea of a generation ship felt like a pointless fantasy when I first encountered it. But as I've spent the last few years speaking with technologists, academics, and policy makers about the hidden dangers of building systems that could reprogram our behavior now and for generations to come, I realized that the generation ship is real. We're on board it right now.

On this planet, our own generation ship, we were once passengers. But now, without any training, we're at the helm. We have built lives for ourselves on this planet that extend far beyond our natural place in this world. And now we are on the verge of reprogramming not only the planet, but one another, for efficiency and profit. We are turning systems loose on the decks of the ship that will fundamentally reshape the behavior of everyone on board, such that they will pass those behaviors on to their progeny, and they might not even realize what they've done. This pattern will repeat itself, and play out over generations in a behavioral and technological cycle. That's what this book is about.

I call this book *The Loop*, but there are, in fact, three loops I want to describe. Think of a small loop nested inside a second, and the second loop nested inside a third. Seeing each one clearly requires examining the others. And I'm going to begin with the smallest loop, the one at the center of everything we do, one that's been spinning for tens of thousands of years, maybe even millions.

The innermost loop is human behavior as we inherited it, the natural tendencies and autopilot functions that evolution gave us. That loop includes everything from racial bias to our inability to accurately perceive risk. That's the first section of this book, because understanding the other loops requires understanding how they rest on this one.

Once we understand that, we can look at the second loop. The second loop is the way that modern forces—consumer technology, capitalism, marketing, politics—have sampled the innermost loop of behavior and reflected those patterns back at us, resulting in everything from our addiction to cigarettes and gambling to systemic racism in real estate and machine learning.

The outermost loop is the one for which this book is named. It's what I worry is coming, a future in which our ancient *and* modern tendencies have been studied, sampled, fed into automated pattern-recognition systems, and sold back to us in servings we will be doubly conditioned to consume without a second thought. It's the loop that has already erased our conscious musical tastes and replaced them with weekly suggestions from Spotify, based on what was served last week. It's the loop that samples our political tendencies and social appetites and puts us together on Facebook with people whose patterns complement our own, such that we become attached to a powerful and all-consuming group identity, and have no patience for people outside of it. It's the loop that looks across all available job applicants and highlights only the ones that match the characteristics of others who've done the job well, so that you, the recruiter, begin to assume hiring can happen in a day or two, rather than weeks, without ever leaving your home. The Loop is a downward tailspin of shrinking choices, supercharged by capitalist efficiency, in which human agency is under threat from irresistible systems packaged for our unconscious acceptance.

It's happening already, but the real threat is what will happen over time. The most powerful and lasting effects will take at least a generation to play out. Think of it this way: we're guided by unconscious tendencies, but we rarely detect it when they are analyzed and played upon. Now throw pattern-recognition technologies and decision-guidance strategies at us. And do it all in a society that doesn't have the long-term sensibilities in our policies and in our programming to recognize and regulate something that will determine the future of the species. That's what this book is about.

I hope to describe here the challenge facing us as we build technology that will alter coming generations of human behavior. Our behavior may be made more efficient, more rewarding, more controlled in the current generation. We may reap immediate economic benefits, and we may be able to do away with difficult and sometimes terrible work in the short term. For every dangerous example I describe here, there will be complex benefits connected to it. But can we look with Darwin's eyes *ahead* at the long-term possibilities of generation after generation experiencing a collapsing cycle of choices, and changing for the worse along the way? Can we somehow predict it, and act to correct it, even when the rewards of doing so won't be felt until you and I and our children are not just long dead, but long forgotten? If we can't, I worry that in a generation or two we'll be an entirely different species—distracted, obedient, helpless to resist the technologies we use to make our choices for us, even when they're the wrong choices.

But I think we can fight The Loop. We just have to stand back, sketch out its shape, recognize the ancient preconditions in all of us that make it possible, spot the modern paths it takes into our lives, and then break its influence on our future.

Jacob Ward
Oakland, California
August 2021

Chapter 1

THE REALITY GAP

JUST AFTER WORLD War I, an Austrian neurologist and psychiatrist named Otto Pötzl began experimenting on a man who'd been shot through the head.

Pötzl, the son of a Viennese journalist, had a journalist's interest in telling big stories through representative characters. This was before the days of evidence-based medicine, back when theories could be based on a single case. And the injured soldier who entered his office was quite a case. This soldier, named Obszut, came through the maelstrom of WWI, in which over a million of his countrymen died in combat, with a single, dramatic injury. His brain was not damaged, nor were his eyes. But a bullet had torn through the connection between the two, and Obszut was blind.

Medical reports from the time show that the shattered men passing through the clinics of Vienna had all sorts of cognitive and perceptual difficulties, and Pötzl undoubtedly knew that many of them were reporting unusual sensory experiences. They were seeing and hearing and feeling things out of order and out of balance. The incoming information that for most of us is generally invisible and incidental—what we glimpse out of the corner of our eyes or what melds into the background of a scene—often overpowered these poor survivors, while the central objects of a room were often invisible to them.

While many of these patients had sprawling, interconnected injuries that made them difficult to study, Obszut's injury was a rare opportunity to experiment on perception in a more formal way. His undamaged eyes could, in theory, take in the light that forms an image. But with the connection to the brain severed, Pötzl wanted to know, would any image find its way into Obszut's consciousness? Doctors at that time were able to enroll their patients in research without clearance from an ethics board, and Pötzl put the remarkably obliging soldier through all sorts of tests.

In a 1917 paper based on their time together, Pötzl wrote that Obszut described having highly sensitive peripheral vision, and at the same time had largely lost the center of his sight. Obszut could, from time to time, even describe the shape of his blind spot, a sort of black orb floating between him and the world. And in the tiny amount of peripheral vision available to him, Obszut also described constant double vision. But the malady on which Pötzl focused was the way images would float into Obszut's consciousness out of sync. Although the soldier often couldn't consciously see something placed in front of him, when that thing was removed details of it would appear in his memory, or even appear in his vision while he was looking at something else. When Pötzl put in front of Obszut a bouquet of flowers with a single piece of asparagus among them, for instance, the poor man couldn't see the asparagus. But when the bouquet was replaced with a second picture, and Obszut worked to catch it in his remaining peripheral vision, he reported a few details of the picture, but with a stalk of asparagus floating in front of it.[1] Somehow the visual information was entering Obszut's eyes, and then, because a bullet had severed the main highway, it was somehow making its way into his brain by another, mysterious road, and arriving at its destination after so much time had passed that the experience felt hallucinatory.

Pötzl wound up describing this concept as "delayed piecemeal delivery into consciousness"—images and sounds and other stimuli somehow appearing in the brain out of order. It was as if the information of the world was arriving in Obszut's head in fragments, like trains full of commuters arriving at a busy train station throughout the morning.

But perhaps this staggered arrival wasn't a flaw unique to this one injured man. The damage to his brain hadn't *caused* these trains to arrive at different times, Pötzl theorized. He surmised that *we all* receive information like this, it's just that those of us with whole, healthy perceptual systems don't have to consciously watch the commuters arrive in our consciousness. We only see the result when they have already finished their haphazard journey to work and are seated at their desks. Our brain doesn't feed us the chaos of the operation as it takes place, Pötzl came to believe. It summarizes everything for us when it's over.

So he began working on the idea that some system in our brains holds back all that stimuli, orders it for us, keeps us from becoming overwhelmed. He called it an abstracting process, and it set him on the idea that there is a world of input we receive and compute but do not consciously experience.

Soon he needed more than just the rare survivors of bullets to the brain. He needed healthy patients, and he wanted to find the moments when most of us, with our undamaged brains, might manifest stimuli we had received unconsciously. He was after what he called "revealing phenomena." And so he began working on dreams. He figured those were places you and I re-experience stimuli we never consciously perceive in our waking lives. So he began flashing images at his patients for less than 10 milliseconds at a time, and asking them to return the next day and tell him about their dreams. And sure enough, over time, the patients came back and reported dreams that happened to include bits and pieces of the stuff he'd flashed at them. Today, researchers still call this the Pötzl phenomenon. He presented the idea to the Vienna Psychoanalytic Society in 1917, and even Sigmund Freud, who famously detested laboratory experiments, was impressed. "The questions raised by Pötzl's experiment," Freud wrote, "go far beyond the sphere of dream interpretation."[2]

Yes, they do.

Before I go pinning too much more on this work, it's important to mention that Pötzl seems to have paid dues as a member of the Nazi Party between 1930 and 1933, officially joined the Nazi Party a second time in Austria sometime between 1941 and 1944, and in 1945, when the war had ended, left his position at the University of Vienna under accusations that he hadn't spoken out publicly against the murder of roughly 14,000 patients at the Steinhof mental hospital. Freud went on to describe Pötzl as an "ambiguous character," not least because Pötzl may have participated in the forced sterilization of thousands of people.[3]

At least one Jewish colleague came to Pötzl's defense. His protégé, the psychiatrist and pioneering neurologist Viktor Frankl, a survivor of brief periods at several work and death camps during the Holocaust, later wrote that he considered Pötzl to have been more or less on the right side of history. "Dr. Poetzl, as a party member, wore a swastika on his lapel but he was far from being anti-Semitic," Frankl declared in his autobiography. "But more than that, he actually helped us sabotage the Nazi orders for euthanasia for the mentally ill."[4]

And so Pötzl is confusing at best, someone who did groundbreaking work and then participated to an unclear extent in the horrors the Nazis perpetrated across Europe. He lived out his days in Vienna and died on April 3, 1962. But his experimental work had a profound effect on generations of researchers.

The reason this book begins with a much-admired if ethically dubious scientist is that in the strange relationship between Pötzl and his unfortunate patient Obszut, I see a formative moment: a first, unsteady step toward what would go on to become a broad understanding today, assembled from thousands of experiments, that reality is the simplest story our brains can cobble together from an overwhelming flood of raw sensory information. And there's something about that finding arising from an era of horror and persecution that is also an appropriate reminder that our ethical footing is usually unsteady. Scientific and technological breakthroughs are often made by people with little or no system for evaluating whether they arrived at their discovery by a moral path, or what dangers their work may pose in the future. As I imagine Pötzl frantically writing down what Obszut was telling him—images the soldier's eye had somehow captured and fed into the brain without any understood pathway between them, discoveries the researcher had made without any of the ethical review a university would demand today, in a nation whose most privileged citizens would go on to commit atrocities on a vast scale—I see the start of a world of research and ethical complication that continues to shape our society.

Today, the science Pötzl helped create has many practitioners, and the discipline has been called everything from decision science to heuristics. But its fundamental finding is this: we believe the story our mind is telling us because we believe that's the only story there is. And in this book, I'm hoping to explain that as we build machines and systems that organize, simplify, and mutate our stories for us, we are just as vulnerable to believing those new tales as well.

SINCE THE 1950S, when scientists began to revisit Pötzl's ideas, researchers have been trying to sort out just how much of reality we actually experience. And they've found that the endless phenomena we observe through our senses—the sweating beer and the job application on the table and the moon rotating across the window and the distant sound of lovers quarreling and the warm breeze through the curtains—are smashed and reordered and rewritten as our brain makes the most efficient sense it can of each scene. Here are just a few examples from a vast canon of these findings.

Let's begin with vision. The study of sight has revealed that what we think we're witnessing—the unfiltered stream of objects and people and events we see—is actually a wild variety of stuff, some of it perceived, some of it remembered, all of it felt in a way we haven't fully defined yet.

In 1992, a pair of cognitive neuroscientists, Mel Goodale and David Milner, wrote a groundbreaking paper that sought to define this notion. It suggested that we have two different ways of seeing an object in front of us: one way involves consciously perceiving the object and making decisions about it, another instantaneously guides the moves we make in relation to that object. They put it more succinctly: "vision for perception and vision for action." Writing in a British psychology journal, they admitted that this "two-streams hypothesis" often made them unpopular at parties: "The most difficult aspect of our ideas for many people to accept has been the notion that what we are consciously seeing is not what is in direct control of our visually guided actions," they wrote. "The idea seems to fly in the face of common sense."[5] Our actions *feel* voluntary, and therefore must be under the direct control of our will, right?

Well, no. Goodale and Milner found, in carefully crafted lab experiments, that our perception is not under direct control of our will. When they put an illusion in front of their subjects, what the subjects reported seeing and what their hands did were often dramatically disconnected. For instance, remember that fun-house illusion in which a face is carved into the wall but looks as if it's protruding out from it? The one where you walk past and the face seems to follow you? Well, when they showed people that illusion, called a "hollow face," the conscious mind was usually fooled, and thought a sculpture was turning its head to track them as they crossed in front of it. (I remember this very thing scaring the bejesus out of me at a haunted house as a child.) But when they asked the same person to flick a small bug-like target off that face, their hands reached *inside* the hollow space that had fooled the eye, past the boundary of the face the conscious mind had seen, and flicked the bug at exactly the right spot. As the researchers describe it, these were "two parallel visual systems—each constructing its own version of reality."

Other senses also seem to be unconsciously received and assembled in the mind as well. The human talent known as stereognosis—the ability to identify an object by touch alone—helps give us incredible manual dexterity. But the processing of texture, weight, and the rest of it seems to take place in the parietal lobe of the brain, a long way from the portion that controls our hands. That may be why stereognosis begins to crumble in patients suffering from Alzheimer's—a disease that affects the parietal lobe—whereas patients with other forms of dementia retain that ability. Your fingertip doesn't experience reality. The brain assembles the things you've touched into its own imaginary rendition of them.

And hearing isn't just a means of identifying a sound and triangulating its position—it also seems to be bound up in other mental tasks, like your brain's ability to remember things in the proper order. Researchers call the idea "auditory scaffolding": that hearing supports the acquisition and retention of things like long lists of words or numbers. Studies of children who are deaf have found that although all their other senses are intact and developmentally they're otherwise on track, they tend to do worse than hearing children do on certain tests of sequential abilities, even ones as simple as tapping their fingers together in a prescribed order. We aren't consciously hearing the raw sounds of the world. The brain is using certain aspects of our hearing to put together its own soundtrack.

So if perception isn't a matter of simply seeing, touching, or hearing reality, what is doing the postproduction work right before it's all shown to us? Maybe memory is involved. The neuroscientist David Eagleman spent a decade experimenting with the mental perception of time. One of Eagleman's experiments would have made even Pötzl spit out his tea.

Eagleman and his graduate student Chess Stetson were looking for a way to test the common report that time "slows down" during brief, dangerous events. In my twenties I was headed downhill on my bike when the car next to me suddenly nosed across the road into a parking space. My bike stopped dead against the car's turned front tire, and I flew into the air across the hood. I distinctly remember having an entire conversation with myself as I traveled through space. "You don't have health insurance! You didn't mail in that COBRA application after leaving that job!"

"Yes, I did, actually. I remember stamping it and mailing it last week."

"Oh yes, you're correct, and that's great news, because it looks like you're about to be pretty badly injured." And then I slammed into the asphalt and slid a few feet, breaking my wrist. I remember sitting up, dazed and bloody, looking at my bent bike. It all felt as if my brain had affected the flow of time for my benefit.

Eagleman and Stetson worked to find out whether the brain does, in fact, slow things down. Neuroscientists had already determined that when the average brain receives two images within one hundred milliseconds of each other, it fuses them together, and we see one image. So Eagleman and Stetson built an enormous wristwatch that displayed random numbers just fast enough—less than one hundred milliseconds apart—that subjects couldn't consciously see them. First, the subjects were shown the numbers in a laboratory setting. They reported that the numbers were just a blur. Then Eagleman and Stetson dropped these people off a crane.

No, really, they marched their subjects to the top of a carnival ride, equipped with a helmet and the big wristwatch, dangled them from a cable, and dropped them fifteen stories into a net.

Here is how Eagleman summed up his findings in a 2009 essay:[6]

> The result? Participants weren't able to read the numbers in free fall any better than in the laboratory. This was not because they closed their eyes or didn't pay attention (we monitored for that) but because they could not, after all, see time in slow motion (or in "bullet time," like Neo in *The Matrix*). Nonetheless, their perception of the elapsed duration itself was greatly affected. We asked them to retrospectively reproduce the duration of their fall using a stopwatch. ("Re-create your free-fall in your mind. Press the stopwatch when you are released, then press it again when you feel yourself hit the net.") Here, consistent with the anecdotal reports, their duration estimates of their own fall were a third greater, on average, than their recreations of the fall of others.
>
> How do we make sense of the fact that participants in free fall reported a duration expansion yet gained no increased discrimination capacities in the time domain during the fall? The answer is that time and memory are tightly linked.

It seems that the shorthand our brains—specifically our memories—normally use in describing an experience to us becomes longhand in an emergency, as the amygdala engages and helps to write out richer, more complete memories of the moment. "In a dire situation, your brain may lay down memories in a way that makes them 'stick' better," Eagleman wrote. "Upon replay, the higher density of data would make the event appear to last longer." This may be why time seems to speed up as we grow older, he theorized. As a child, every bus ride is a new experience, and so the memories are intricately crafted. As we get older, the brain says "yeah, yeah, a bus, I know how this one goes," and the memory is recorded in a more slapdash way.

Our experience of the world, from car accidents to optical illusions to the grip of a stranger's handshake, *feels* real to us, in such a consistent way among humans that we can describe the events to one another and agree we've arrived at an accurate retelling of the moment. But the primary thing we have in common is an onboard perception system that curates reality for us. And that mediating layer is fertile soil for the forces we're going to be describing in this book. Why? Because as we've already seen, and will continue to see in the next few chapters, the human brain is built to accept what it's told, especially if what it's told conforms to our expectations and saves us tedious mental work.

NOW LET'S THINK back to the lab where Pötzl put asparagus in front of Obszut. At that time, it was clear to Pötzl only that some sort of mediating layer was doing the work of organizing the soldier's inbound perceptions. But now, more than a century later, researchers are working with similar patients and injuries, and they're finding that our brains are not closed systems. They don't just unconsciously assemble the sights and sounds around us, they also absorb emotions—and *transmit* them—without our conscious awareness.

I met Beatrice de Gelder in New York City while filming a documentary series. She splits her time between the United States and the Netherlands, where she runs a lab at Tilburg University as a professor of cognitive neuroscience. De Gelder built a second career studying the brain after a prior one in philosophy. She also edits the respected journal *Frontiers in Emotion Science*. And yet she says her work, which centers on the idea that vast unconscious mechanisms move visual information into our brains without our realizing it's happening, has always been considered fringe.

Since the late 1990s, when she first began working on the idea that sight is much more than just what we consciously see, she has encountered passionate resistance to it. "There is a bias in human visual studies to focus on the visual route that is associated with consciousness," she explained in accented, perfect English. "I mean, that's what people have always been working on. And only a minority of people have looked at something else."

De Gelder knew that in the early twentieth century, researchers like Pötzl had their choice of ideal patients in which to study what's now known as cortical blindness. After World War I, she told me, "those gun wounds yielded very clean cases of young people that had a specific loss of function." In the modern world, when researchers have been eager to study sight by looking at those robbed of it, their subjects are often compromised in other ways, because when something like a car accident or a stroke severs the visual cortex it also inflicts all sorts of secondary damage on the body that muddies the waters for researchers.

De Gelder's work is possible because she has found her own category of people blinded in a very specific and measurable way: people in whom lesions have formed on the visual cortex, severing them the way a bullet might. These are often people who have experienced a relatively minor stroke that left them blind. But like Obszut, their eyes and brains were intact. The connection is the only missing piece.

In a 1999 paper,[7] de Gelder described working with one patient, codenamed GY to protect his privacy, with whom she attempted several experiments having to do with ventriloquism. Nothing had produced results, and so she and her lab assistants, eager to make the most of the day—"such patients are precious, and by the way, you pay them well for their time"—began playing with putting joyful and fearful faces directly into his blind spot, and asking him to guess which emotion was on display. GY managed to guess the correct answer roughly 75 percent of the time, far better than chance, and de Gelder wondered whether he was in fact seeing the emotion in some unconscious way, the information traveling from eyes to brain via a route other than the disconnected visual cortex, the way Obszut's asparagus did.

That experience set de Gelder on a search for other people experiencing what she came to refer to as "blindsight"—functional eyes and functional brain, disconnected from one another—and eventually she found herself with another patient, codenamed TN, and with more time to kill. As a day of experiments wound down, she and her graduate students decided to assemble a sort of obstacle course in the hallway of their department, and asked TN to hand them his walking stick and make his way straight down it, without telling him that all sorts of objects—trash cans, stools—were in the way. He hesitated, perhaps sensing that they weren't telling him something, and asked for reassurance that they wouldn't let him hurt himself. They told him he'd be fine, and before they knew it, he turned and traversed the jumbled passage without touching anything. It was uncanny, de Gelder told me. As he reached the first obstacle, TN stopped and picked his way around it, and then another, turning this way and that to avoid the items in his path.

He never faltered, and he arrived at the far end of the hall easily, as everyone gasped in amazement. Everyone except his wife. Turning calmly to de Gelder, she told her that this was normal. He makes his way around the house without any trouble, she explained.

The researchers crowded around the man, asking him how he had done it. "Done what?" he asked them. He had no idea he'd avoided anything. As far as he was concerned, he'd taken a straight walk down a hallway.

What's so amazing about this is that it not only suggests TN was somehow getting visual information about the obstacles into his brain without the benefit of his visual cortex—which means the information was moving via some other, undiscovered pathway—but there was also likely some unconscious process that was absorbing the information, and issuing instructions to the man's motor functions, causing him to pick his way around the obstacles without realizing it. "We were very, very excited," de Gelder recalls. "We actually made him do it again"—this time on video—"and we never thought it would work, or I would have made sure the recording was a bit better."

The video is a wonder to watch: a researcher holds TN by the shoulders to point him in the right direction, and then, when TN is released, he walks up to the jumble of wastebaskets and chairs and makes his way through it all just as I might have. He stops and edges around things, he slides sideways when he needs to. He acts like a man who is seeing what's in front of him.

Soon de Gelder was working on determining what other things blindsighted people like TN can see without realizing it, and she returned to emotions. In one experiment, the subjects were shown pictures of emotionally charged faces—people who are obviously angry, or sad, or smiling. And although the subjects told de Gelder and her students—probably with some irritation—that they only saw a blank screen ("because hello, I'm blind!"), electrodes attached to their faces detected tiny, micro-muscular activity: electrical signals of facial expressions indicative of the same emotion. In other words, their faces unconsciously mirrored the expression they could not consciously see. Smiles inspired smiles. Frowns inspired frowns. Evolution, it seems, may have given us a second, unconscious pathway for seeing emotions. And not only that, it somehow gave us the gift of automatically *passing them on* to our fellow humans.

De Gelder's findings are still controversial. As she describes it, "People in mainstream visual science don't work on emotions." But she has seen greater acceptance since her 1999 paper documenting GY's emotion-guessing abilities. "Things are rapidly changing, especially the last 10 years," she said. "But emotion is still not an entirely respectable topic."

Nonetheless, de Gelder says she believes there's an entire language of gestures, facial expressions, and movements that you and I unconsciously use to communicate visually. "Sensitivity to the direction of the gaze is a big one," she says. In her tests, humans have been shown to unconsciously and instantaneously register the moment someone breaks eye contact, and in which direction. Even five degrees of shift, and our brain knows it. Why do we have such a talent for following the gaze of others? "Deviation from direct gaze tells you where to look," she says. "And if you compare the speed of reading someone else's gaze to how long it takes for them to instruct you with their voice where to look? Well, it's a great method for saving time."

For de Gelder, the evolutionary logic is clear. "The brain is in charge of pretty much everything we do, right? Finding food, staying warm, finding shelter, staying away from harm, and all that stuff takes resources," she told me. And if it can save time and effort by taking unconscious clues from others and acting on them without having to bother your conscious mind, all the better. "It gets you reliable information at a lesser price than having to figure it all out yourself," she said. "If you can trust the expression of fear on somebody else's face, you can gain a lot of time and reliability for your actions."

Between her work and that of many others, there is growing evidence of our unconscious ability to absorb, act on, and transmit logistical and emotional information. The result is a constructed, and perhaps even shared, story our brains tell about reality. In effect, our conscious minds are experiencing our lives secondhand.

Whole fields of research are now devoted to this phenomenon, the mental gap between our perception of what's going on and what's actually going on. Experts have learned to measure that gap. They've begun to discover the patterns we follow as we make decisions inside it. And in the process, they're also finding that our modern world is powerfully shaped by unconscious habits that follow us through our lives.

The efficiency and economy of our mind, the ways it packages the world up for us, the ways it seemingly communicates with other human minds, increasingly suggest there is a grand pattern of invisible guidance we follow through our lives, while believing we are in fact making our own decisions. And if science is only just beginning to decode these systems, systems that have presumably been steering us from place to place for as long as our species has been in its modern form, that also suggests we have no real conscious sense at all of just how easy it is for us to fall into patterns beyond our control, or which outside influences may be exerting irresistible control over us. If you and I were asked to rank the forces that have the greatest unconscious influence on what we choose to do in life, we'd have no idea where to begin. And yet as research like de Gelder's shows, there is a whole apparatus inside us that is receiving, ranking, and acting on those influences throughout our lives. All of that creates an enormous opportunity for taking advantage of—and even directing—human behavior.

Why directing? Because discoveries like de Gelder's aren't just being used in a medical or academic sense to try to close the reality gap. Unconscious choice is still very much in its infancy as a scientific field, but that hasn't held back efforts to turn its findings against us. People outside the sciences—specifically in the world of business and politics—have been catching on to the idea that humans have an unconscious "code" by which they make choices. More and more, companies and political operatives have sought to use that code to shift our behavior. And while the first wave of that sort of research was fairly primitive, there are now whole courses taught at top universities around the world about how to use these findings to persuade customers and sway voters, as we'll examine in the coming chapters.

Now that we've seen how powerfully our unconscious brains shape our lives, let's look more deeply at how they tell us what to do. Because if we don't familiarize ourselves with the mechanisms of our brains, we'll not only be vulnerable to those who prey on us, we'll be blind to the effects. We'll be caught in an endless, cluttered hallway like the one TN traversed, turning this way and that in the thrall of unconscious processes we don't realize are guiding us.

Chapter 2

ILLUSIONS

IT MAY SEEM as if a book about the influence of cutting-edge technology should spend most of its pages on that technology: who makes it, where it's headed. That's all coming, but to really see the threats here, we need to fully grasp the implications of the science underpinning them. What persuaded me to write this book wasn't the technology. It was the breakthrough science that has revealed our most automatic habits. Our unconscious tendencies are the control surfaces by which technology will shape our lives. And for me, the threat comes into sharpest focus the more I learn about our psychological aversion to seeing these tendencies in ourselves, and about the business and cultural forces working to convince us we're making independent choices when we're doing the opposite.

Seeing ourselves clearly is the first step in understanding the threat that The Loop poses. It's the first loop that supports the second and third. And so the next few chapters of this book are going to lay out a series of unconscious tendencies—internal mechanisms that click along in the background, outside of our awareness—and the scientific efforts that have slowly identified and measured them. Then, and only then, can we understand the scale and power of what's coming.

THE CURRENCY OF science is the published paper. A scientist or group of scientists writes out a hypothesis and its supporting evidence, other scientists critique the work as part of "peer review," and a journal releases the revised result to the world. Then, if all goes well, other scientists try to do the same sort of experiments and publish the same conclusions to confirm the findings. A scientific career is often evaluated, fairly or not, by the sum total of this published work. And the pace of science remains very slow as a result, even as the world speeds up around it. Scientists can often remain at odds with one another for years over a finding that in the meantime becomes the functional basis for whole industries.

Scientists today publish roughly 2.5 million papers each year. Presumably they all hope each paper will make an impact. But only a tiny percentage ever do.

The psychologists Daniel Kahneman and Amos Tversky wrote seven papers together between 1971 and 1979. They blew a gaping hole in their field with each one, and while their work is still being tested and replicated at the pace of science, their findings are already a tent pole of a whole industry of behavioral guidance. Working together in an extraordinarily tight and symbiotic professional partnership, they identified a series of commonly held biases—unconscious human preferences that manifest themselves under pressure and in moments of uncertainty, preferences that often produce bizarre and irrational decisions. And they went a very long way toward demonstrating that what they'd discovered afflicts us all.

By putting various logic puzzles, gambling propositions, and guessing games in front of unwitting subjects, they worked out the boundaries of common heuristics, or the rules people use to make decisions. The papers they authored together upended decades of psychological and economic theory. In a single paper, "Judgment Under Uncertainty: Heuristics and Biases," they laid out a dozen systematic biases and a trio of heuristics—calling them representativeness, availability, and anchoring—that form a trinity of human weirdness. That one paper set off a decades-long inquiry that is still reshaping our understanding of who we are.

The three heuristics in the 1974 paper might seem obvious, behaviors so familiar to you that it feels silly to hear them named and analyzed. But with a little squinting we can see just how well defined and powerful they are. The first, representativeness, is the tendency of human beings to assume that if something has the qualities they associate with a certain category, it must belong to that category. It's a simple concept, but its implications were and are explosive. Kahneman and Tversky tricked their subjects into revealing their assumptions about certain categories of people, objects, and occurrences. For instance, they asked students to rate the likelihood that Steve, a shy and tidy man, was either a farmer, salesman, airline pilot, librarian, or physician. Librarian regularly topped the list.

"This approach to the judgment of probability leads to serious errors," they wrote, "because similarity, or representativeness is not influenced by several factors that should affect judgments of probability." People simply weren't going about the task of ranking Steve's likely occupations in a rational way. There are vastly more farmers than librarians, and yet the fact that his personal qualities reminded them of librarians (and mistakenly, as librarians spend most of their time dealing with the public) caused the subjects to make wild assumptions about how Steve spends his days.

Kahneman and Tversky went on in the paper to describe how our vulnerability to representativeness also means we often make mistakes about the nature of chance, the effects of sample size, whether we have reliable information, and whether something can be predicted at all. They fundamentally destabilized our self-image as a rational species. And that was just the fourth page of the paper.

Next, they turned to the availability heuristic, in which "people assess the frequency of a class or the probability of an event by the ease with which instances of occurrences can be brought to mind." Our ability to remember something alters how common we consider that thing to be. Why is that a problem? "Availability"—our ability to think of an example—"is affected by factors other than frequency and probability," they wrote. This was another seemingly simple observation, again explosively important. For one, just because an event stands out in the mind doesn't mean it's a common occurrence. Let's say I'm planning a vacation, but I just witnessed a hotel fire in the city where I live. I might decide that hotels are too prone to fire, and that I won't be staying in one this trip. We make this sort of decision based on perceived probability all the time, especially when the risk is something vividly terrible, like fire. But we're misreading the odds, it turns out. Seeing a fire makes people think fires are more likely than does reading about a fire in the paper, Kahneman and Tversky pointed out, but the emotional impact of watching the flames spread from the smashed windows to the roof doesn't actually change how often fires take place. It just plants it in the mind, and shapes the choices we make about it. What else makes a memory stick in this way? Some things are simply easier to memorize, like familiar destination names, as opposed to unfamiliar names you can't instinctively spell. We fixate on hypothetical scenarios that are easy to imagine (missing the plane, forgetting a jacket). And we connect things that are entirely unrelated, but feel linked, like forgetting an umbrella and winding up in a long weekend of endless rain. All of this seems to us like a series of reasonable assumptions about probability and risk and reward. But a lot of it is illusion. The availability of certain events and factors in our memory makes them leap to mind, and feel as if they're more likely than they are.

Finally, the paper describes the anchoring heuristic. This one is largely mechanical—a misaligned carpet on which we consistently trip—and as a result, at least to my mind, it's the hardest to accept as being true. Maybe I just don't like being told I have a tic I wasn't aware of. And yet here it is: anchoring is the simple tendency to "make estimates by starting from an initial value that is adjusted to yield the final answer." We lock on to an early idea, often at random, and then adjust it as we go. Trouble is, once we've locked on, we don't adequately adjust, even in the face of overwhelming evidence that we're wrong.

The experiments Kahneman and Tversky used to establish the anchoring heuristic were like little bar bets where the right answers win you a free round of drinks. One of them asked subjects to estimate what percentage of the United Nations were located in Africa, presumably on the assumption that very few Western participants in 1974 would know the answer. (It was roughly 25 percent at the time.) But first, the idea was to see if the experiment could prime the subjects to latch on to a random number, so they were shown a random number between 0 and 100, *Wheel of Fortune* style, on a spinning disc. It's ridiculous—they were asked to guess a number they didn't know, after viewing a number they knew to be random. And yet their minds anchored to the random number: people shown a larger number tended to estimate a greater number of African nations, and subjects shown a smaller number estimated a lesser number. Even when the participants were *paid for accuracy*—Take your time! Really *think* about it!—the anchoring bias didn't go away.

Kahneman and Tversky pointed out that for this particular parlor trick to work, the subjects had to lack the necessary information or computational resources to make an informed choice. Basically, they had to be guessing and under pressure. If they were, people made the same systematic mistakes. They locked onto the value of the first numbers they encountered, and they expressed an irrational confidence in their estimates despite not having enough information. And for our purposes, it's important to also recognize that these people, presented with a semi-technical question about statehood, watching a spinning wheel of numbers—in other words, faced with a system they did not understand—were powerfully influenced to make choices they *did not consciously understand*. We will see that same tendency again and again in this book.

The paper, as if to reassure us about the utility of the embarrassing tendencies it describes, explains that while these heuristics create in us outrageous cognitive vulnerabilities, they are, in fact, wildly useful overall. They're quick and dirty shortcuts by which humans make efficient decisions, decisions that helped our ancestors stay alive. Instead of having to process every minor detail of the environment, efficient mental systems—for making quick judgments about one another, for recalling what's important, for guessing at numbers—saved us valuable time in choosing allies, assessing risk, finding food. But these shortcuts are also clearly dangerous. "These heuristics are highly economical and usually effective, but they lead to systematic and predictable errors," they wrote.

Kahneman and Tversky had identified three handy ways we make short work of the world's complications, but they also knew the three heuristics posed a real danger to our ability to make good choices in the modern world. Representativeness, availability, and anchoring would make us prone to believe the most familiar, most memorable, most recent story we hear. As they'd written in a 1973 paper, "The production of a compelling scenario is likely to constrain future thinking."

Their work was a revelation. The findings, described mathematically, stood up over time—a key test of their scientific merit. (Anchoring, in a replicability study in 1995, at a time when other areas of psychology were being torn apart as junk science, was in fact shown to be *stronger* than they'd originally measured.) And they were such original ideas that they set off a cascade of related findings.

The graduate students working under or around Kahneman and Tversky felt enormous pressure to contribute. In a short history of his field, the psychologist Baruch Fischhoff, who was among those students, wrote that "the challenge for the participating graduate students was to find a heuristic to call one's own or to find a way to elaborate one of the three 'classics.'"

Fischhoff, a native of Detroit, moved to a kibbutz in Israel after college, and he intended to spend his life farming. However, he told me, "my wife is the more sensible of us. And she couldn't take it anymore." As a college math major, Fischhoff had taken psychology because students had to take a certain number of classes outside their major, and because the psychology department's schedule fit his 4 p.m. to midnight work shift at Great Lake Lanes, the local bowling alley. When he decided he preferred life as a married academic, rather than as a divorced farmer, his unusual background in math and psychology earned him a place at Hebrew University of Jerusalem, studying for his PhD under Amos Tversky.

"I was in a very progressive, activist part of my life," Fischhoff recalls. "I was reading a lot of history, I was interested in politics, and the subject of hindsight was something I'd been thinking about." Paul Meehl, a University of Minnesota professor of psychology, had just published, in 1973, a paper titled "Why I Do Not Attend Case Conferences," a twelve-point critique of the tendencies of experts gathered at professional meetings to make a hash of their subject after endless arguing and schmoozing. "I thought I could build some science out of that idea," Fischhoff says. "So I came up with a handful of experiments, and they stood up ridiculously well over time."

What Fischhoff established was another systematic error: the hindsight bias. In his paper, "Hindsight ≠ Foresight: The Effect of Outcome Knowledge on Judgment Under Uncertainty," Fischhoff coined the term "creeping determinism," which he described as "the tendency to perceive reported outcomes as having been relatively inevitable."

In the study, when Fischhoff described the outcome of a historical situation to participants—Nixon's visit to China, a military engagement between British and Nepalese forces—they said that the outcome was the most probable, and they also perceived the odds of it happening as roughly double what people who didn't know the outcome would estimate. What had happened in the past looked to them as if it were somehow preordained. They were biased by hindsight.

And beyond that, they were largely blind to that bias. In a second portion of the study Fischhoff would ask them to predict an outcome without knowing what had happened. And once they'd done so, participants interviewed later turned out to not only forget what outcome they'd predicted, they thought too highly of their judgment in retrospect. In fact, once they learned what had gone on to happen, they tended to remember having predicted something close to it, even when that wasn't true. They could not retrieve the psychological experience of lacking accurate, backward-looking information. Once they had that information, they were under the illusion that they'd always had it. Their remembered selves had better judgment than their actual selves.

"Subjects remembered having given higher probabilities than they actually had to events believed to have occurred," Fischhoff wrote, "and lower probabilities to events that hadn't occurred."[1] They made inaccurately confident statements about the past, and *they misremembered what they had predicted at the time.* Fischhoff had discovered that we humans are not the insightful historians we think we are, concluding that "the feeling that we understand what the past was all about may prevent us from learning anything from it."

The cascading effects of Kahneman and Tversky's work continued for decades, further strengthening a new scientific understanding of the weird and irrational ways humans make decisions. Paul Slovic, who studied with Fischhoff in Jerusalem—and was essentially marooned there with him while Kahneman and Tversky served in the military from 1973 to 1974—continued to work in the field, and identified a new, incredibly powerful bias in 2000. And in doing so, he discovered that academics weren't the only ones studying heuristics.

Slovic began a sort of think tank consultancy called Decision Research in 1976. From its headquarters in Eugene, Oregon, he and his team (which came to include Fischhoff, psychologist Sarah Lichtenstein, and many others influenced by Kahneman and Tversky) worked on questions of risk, influence, and uncertainty, mostly in the public sector. Slovic says he first understood the power of emotions to drive our behavior during one of his first projects there, when he was asked to study the work of a particular group of researchers. "They were decades ahead of the rest of us," he says. But they weren't an academic department or a think tank. They worked for cigarette companies.

"These folks had figured out that it does the company no good to try to sell cigarettes based on a comparative analysis," Slovic remembers. Marketing a cigarette as being somehow the better choice than another brand was an inferior strategy to simply packing the idea of the brand with emotional content, they'd found. "Calling the cigarette 'Kool,' and using phrases that carry emotional meaning, like 'alive with pleasure'—that's the stuff that got people to buy it," Slovic says.

At the same time, he also worked on risk. As he looked at topic after topic—smoking, nuclear power, genetic experimentation—Slovic found that people were not rational in how they assessed the chance of something bad happening. In 1987, he published a study, "Perception of Risk," that took a shot at identifying the differences between how you and I think about risk—a fog of perception and uncertainty—and how experts rationally evaluate the same dangers.

But he thought he wasn't quite there. "This is something that evolved over time," he said. The assessments we make of everything from smoking to world-ending weapons seemed to Slovic to be tied to more than just our intellect. He suspected that our emotions were involved, and not just as a byproduct of what we decide. He thought perhaps feelings were an assessment system of their own. "I found myself playing with this notion that images and associations connected with feelings became connected with motivations for behavior."

Slovic's work plugs into a long tradition of studying the mental role of emotions. In 1884 and 1885, two academics, William James and Carl Lange, working independently, arrived at the idea that emotions are the mind's way of processing the physiological experience of a high-stakes situation. They were each seeking to explain the sheer efficiency of emotions like fear. The James-Lange theory, as it came to be called once they found one another, has been most often summarized using a hypothetical bear: We don't run because we're afraid of the bear. We're afraid of the bear because we run. The body, in a sense, is responding first, and then the brain builds a narrative that helps package up the experience so that we can understand it. Psychologists object to the bear analogy as an oversimplification, but it gets at the fundamental idea. The emotions, James and Lange believed, are a way of receiving complex information packaged in a format we know what to do with. James, famous for broad proclamations, made an especially broad one in an 1899 publication:

> Ninety-nine hundredths or, possibly, nine hundred and ninety-nine thousandths of our activity is purely auto-matic and habitual, from our rising in the morning to our lying down each night.[2]

Modern researchers built on this theory. In 1952, neuroscientist Paul MacLean coined the term "limbic system" to describe the "visceral brain" we use to process emotions. In 1980, University of Michigan psychologist Robert Zajonc ran experiments using hidden sounds in an audio recording. He showed that people displayed an emotional reaction to those tones, although they couldn't consciously remember them. His measurements suggested that emotions may be, in fact, steering the ship, and consciousness is just a passenger. Maybe we really do only experience fear once it has us up and running from the bear. Zajonc wrote in 1980 that "we sometimes delude ourselves that we proceed in a rational manner and weigh all the pros and cons of the various alternatives. But this is probably seldom the actual case. Quite often 'I decided in favor of X' is no more than 'I liked X.'"

Building on this work and what he'd seen studying around risk, Slovic now wanted to try to prove that emotions truly guide our decisions. In 2000, Slovic and colleagues from Decision Research and the Islamic University in Saudi Arabia published a paper that named a new bias: the affect heuristic.

In "The Affect Heuristic in Judgments of Risks and Benefits," Slovic and his colleagues wrote that our emotions stood alongside availability and representativeness as an unconscious system for making important decisions. (Anchoring, meanwhile, was something psychologists had found was real, but for which they couldn't quite articulate an evolutionary purpose. It's often left out of later work.) The subjects' positive or negative emotions about the benefits of, say, nuclear power, the paper claimed, tended to change their estimates of the risks of nuclear power. As Slovic described it in a 2006 paper with psychologist Ellen Peters, "People judge a risk not only by what they think about it but also by how they feel about it. If their feelings toward an activity are favorable, they tend to judge the risks as low and the benefits as high; if their feelings toward the activity are unfavorable, they tend to make the opposite judgment—high risk and low benefit."[3] That idea was as relevant in 2006, when people signed up for variable-rate mortgages and banks packaged them as investments, as it was in 2020 as otherwise careful people chose to forgo masks during a pandemic and risked dying on a ventilator just to spend time with loved ones at a party. Kahneman once described the affect heuristic as "probably the most important development in the study of judgment heuristics in the past few decades."

Why do we need to understand these seminal discoveries by behavioral psychologists in order to understand The Loop, this generational spiral of shrinking choices? It's not just because it's important to understand and accept that we misinterpret information, we make inaccurate assumptions, and we get confused. We've always known this is part of being human. But Kahneman, Tversky, Fischhoff, Slovic, and dozens of others are at the beginning of understanding the granular *patterns* of human irrationality. That mediating layer we talked about, the one that packages reality for us? It turns out to have its own foibles and allergies. And those vulnerabilities are just as important as the system's strengths when it comes to understanding who we are, and how we may be manipulated in future.

The new difficulty we face as a species is understanding just how powerfully these vulnerabilities play on us at exactly the moment when we think we're actually conquering our irrationality—building automated systems to help us with risk assessment, pattern recognition, and tedious *work*. And not only that, we're building a vast industry of robots, AI, and other assistive technologies that our brains won't be able to *help* but lean on—long before those systems are ready to support our weight.

As Kahneman and Tversky and so many other researchers have established, our brains are shortcut machines, desperate to hand off difficult cognitive tasks. We're about to see how this happens—and why it makes possible a whole industry of decision-shaping technology that we cannot see or feel, and yet we will be powerfully dependent upon.

Chapter 3

TWO SYSTEMS

IN 2013, I stood under the burning sun in Homestead, Florida, watching robots fall over.

They were attempting to navigate an obstacle course on a sun-bleached motor speedway where I and dozens of other reporters from around the world squinted in the heat, even through our sunglasses. The assembled machines were intricate contraptions. One of them could scurry around on all fours across a field of rubble, then rise onto its hind legs to grasp at a door handle. (That one managed to play on both my fear of insects *and* my fear of home invasion.) Another had a sort of stooped, careful stance that reminded me of my grandfather shooting pool. A third had a sleek white and gold exoskeleton that could someday qualify it to check IDs outside a nightclub. They all *looked* very impressive. The products of the best minds in civilian, military, and academic robotics from all over the world, these creations were part of a tournament called the DARPA Robotics Challenge. I'd flown in with a camera crew to cover the event, promising my bosses at Al Jazeera that this would be made-for-TV mechanical drama at its best. But that morning I realized I was going to have to apologize, because an early task was crushing every robot.

They were all trying to get up a ladder.

DARPA is the Defense Advanced Research Projects Agency, a think tank for the Defense Department. It exists to pull far-out ambitions into reality. An earlier iteration of the agency pursued a networked information system that could survive a nuclear attack and wound up inventing the Internet. These days, in addition to offering hundreds of millions of dollars in grant money, it also holds highly lucrative competitions. This particular competition in Florida was an annual competition, a yearly federal goosing of the robotics industry into producing an all-in-one robot helper for disaster work. (Although of course a robot capable of these tasks could have other, more military purposes as well.) This same competition format previously produced a functional self-driving car, after several years during which the entrants careened off roadways and crashed into barriers. But this year it didn't reach such heights with the robotic assistants that DARPA imagined someday putting out fires, handling nuclear material, or carrying lunch across a minefield.

The trouble was that these robots were faced with human tasks, in human environments. To an outside observer like me, who knew next to nothing about building a robot, the jobs looked easy. Open a door. Use a drill. Get out of a car. The ladder, if anything, should have been the simplest. To the human eye, a ladder consists of a perfectly reasonable number of variables. The handholds are smooth and inviting, the rungs are consistently spaced and wide enough to comfortably hold a foot, and when balanced right, it's very sturdy. This particular one, built to hold thousands of pounds of robot, was more like the stairs one might ascend to give a speech—a kid could easily get up it—and yet it absolutely murdered these robots.

The problem is data. Even though the ladder *looks* simple, the incoming variables are in fact overwhelming. A robot approaching the staircase is presented with reams and reams of unfamiliar information. And it has to consider each item individually. Where is the first rung? Is it visually separated from the ground? How about the second rung? Can one infer from the relation between rungs one and two what the angle of the whole thing is going to be? Does the railing follow the same angle? And where does it end? The tiniest interaction with the ladder turns out to be incredibly complicated. And no one task can be done *instinctively* by a robot, as you and I do so many things. The robot cannot unconsciously throw a foot the same distance each time, feeling for the foothold with a toe, and the actuators that keep it in balance don't coordinate automatically with the signals coming from its lower extremities, as mine do. The robots have to consider each and every factor, no matter how small. It all has to be processed *consciously*, if that's the word. And they just didn't have enough conscious bandwidth to handle the situation. Time and again, the robots in Florida went over—some backward, some sideways, some headfirst into the rungs—and a helmeted member of the robotics team in question had to save them at the last moment, desperately hoisting the falling robot off the stairs on a belay rope, everyone on the team cursing into the sky or at their own feet as their mechanical champion dangled in the air.

Even though I'm arguing that we should be worried about the advanced state of our technology, and these robots were pathetically bad at ladder-climbing, I'm telling you about their laughable failings for two reasons.

First, they represent the grand ambitions of technologists to build sophisticated stand-ins for the human mind and body. In two decades covering technology as a reporter, I have over and over again been shown technology whose creators talk about replacing or supplementing humans in all manner of tasks. But the robots of the DARPA Challenge, and their painful efforts to stand in for us, are the gap between what we want technology to do, and what it can actually accomplish. These robots are fascinating and frightening, but they are not at all ready to replace humans, no matter how desperately we might want to hand one a fire hose and tell it to start saving lives. It may be that someday they *will* fix broken screen doors and extinguish fires for us. But our desire to give them those jobs—and our unconscious willingness to believe that they *can* do them—doesn't match the robots themselves. I went to Florida expecting miraculous artificial beings. I left Florida with my expectations reset. I've tried to carry those expectations with me ever since.

The other reason to consider these robots here is that their failure to do what we do tells us something fundamental *about* what we do. Just as Kahneman and Tversky showed that we process uncertainty in an unconscious way, it turns out we process the rungs of the ladder, a soup spoon, and our shoelaces outside our consciousness. The ability to see and manipulate these objects takes place outside our waking minds. It's a system that has kept us alive for millions of years because it takes the load off our consciousness, freeing it up to do more important things, like invent new ladders and spoons and shoe closures. It's an extraordinary gift. And it's important to understand it, to understand that we haven't yet built a stand-in for it, and to understand that we are nonetheless geared to unconsciously offload difficult cognitive tasks to automatic systems, whether they be our emotions or a collection of shiny but unreliable robots, even when they're not up to the job.

WHY DO WE do this? Why do we have so many automatic tendencies, the kind that let us climb a ladder without much thought? How do our mental systems divide the grunt work from the bigger, thinkier stuff? In 2000, the psychologists Keith Stanovich and Richard West published a paper that looked back across three decades of people variously kicking around dual-process theories—the widely accepted idea that we have both an unconscious and a conscious mind. Researchers had variously argued that one mind formulates biases, whereas the other doles out analysis; or that one makes associations, whereas the other obeys rules. But Stanovich and West waved away the small differences. They pointed out that while the individual theories put forth by a dozen papers "do not always match exactly, nevertheless there are clear family resemblances." And so they cut through all the disagreement and simply declared the existence of a System 1 and a System 2.

System 1, they wrote, "is characterized as automatic, largely unconscious, and relatively undemanding of computational capacity," whereas System 2 handles "controlled processing" and "analytic intelligence."

Basically, System 1 makes your snap judgments for you, without demanding conscious attention or consuming much energy. System 2 makes your careful, creative, rational decisions, which take valuable time and fuel to process.

System 1, in other words, is what gets you up the ladder. Its process of making decisions is an unconscious form of mental activity that takes in information, decides what to do with it, and enacts a response. System 1 balances the easel, unscrews the caps of the paint tubes without dropping them, and holds the brush. System 2 is the part of your mind that's meanwhile free to think, "Hmm, what shall I paint?"

In 2002, after a career working variously alone and with Amos Tversky, who died of cancer in 1996, Daniel Kahneman won the Nobel Prize. In his speech to the committee (which I would have spent simply congratulating myself), he described his new interest in what Stanovich and West were talking about. He began by pointing out that he and Tversky had always imagined the biases they were identifying sprang from a place "between the automatic operations of perception and the deliberate operations of reasoning." Now he wanted to explore that notion formally, and so he was setting out to explore System 1 and System 2.

In a 2003 paper, "A Perspective on Judgment and Choice," based on his Nobel speech, Kahneman pointed out that a failure to make a rational choice isn't just System 1's mistake. It's an error in both systems: "System 1, which generates the error, and System 2, which fails to detect and correct it."[1]

Think of System 2 as a watchful older sibling charged with keeping a younger sibling out of trouble. It's a tough job. For one thing, as Kahneman pointed out had been established by other researchers, "most behavior is intuitive, skilled, unproblematic, and successful." In other words, the younger sibling does most of the work around the house, and by and large does it the right way.

And the older sibling, System 2, easily loses focus. System 2 basically can't do the job except under ideal circumstances. If it's under stress, or distracted by another task, if it's too tired, well, the younger sibling is left to just carry on.

And even under ideal circumstances, unless System 1 is making an obvious mess of things—pouring gasoline into a blender or something—System 2 doesn't have the cues it needs to recognize it's time to step in. When System 1 looks like it knows what it's doing, the smarter older sibling stays out of the way. Researchers had already found that most humans have roughly the same System 1, almost no matter their cognitive abilities, and as Stanovich and West pointed out in their paper, those unconscious talents "show little relation to analytic intelligence." The younger sibling in each of us, who does most of the work in our lives, doesn't draw on the "intelligence" we work so hard to cultivate. Being intelligent doesn't necessarily help, because, as Kahneman concluded, unless System 1 makes an obvious mistake at a time when System 2 is available to notice, "there is no opportunity for intelligence or sophistication to manifest itself." While we'd obviously want System 2 to step in and veto the worst decisions of System 1, Kahneman suggested that most of the time System 2 either doesn't get involved, or ratifies System 1's decisions with a handful of small tweaks, often ineffectual or counterproductive ones. ("This ride on the back of my friend's motorcycle should be fine. I'll just close my eyes when it gets scary.") This is what we do when we rationalize instinctive choices with grand narratives about "going with our gut," or mask our unconscious decision to buy something unnecessary, like an expensive bottle of wine, with a seemingly rational spiel about the value we're getting at that price. Sometimes System 2 simply serves to fool us into better trusting the path System 1 has unthinkingly taken.

But before I go on badmouthing System 1, let's talk about the good it can do.

Slovic describes the power of our instincts this way: "Our experiential system is extremely sophisticated, and accurate. It's extraordinary. When you think about human societies and cultures interacting in complex ways with technology and weapons and the rest, and the fact that most of the time we're doing this with our experiential system, and it basically works pretty well." He laughs. "So much of it gets us through our day, and gets us where we want to go."

Our System 1 evolved to instantaneously spot calories and weigh the value of those calories against the risks of climbing a tree or tracking prey to get them. It evolved to distinguish kin from strangers without our having to think about it at all. It's a rapid-action guidance system, built for survival. Even though *our* day—checking social media, talking with our kids over breakfast, sitting down to work, making professional decisions, political decisions, financial decisions—is not the one System 1 evolved to get us through, we've carried that system with us into our twenty-first-century lives. And, suited for it or not, its ability to handle most of our decision-making for us is still saving us an incredible amount of processing power.

It's a good thing, too, because our brains are expensive to run. Despite making up only 2 percent of our body weight, the brain consumes more than 20 percent of our energy. (The brain, kidney, liver, heart, and gastrointestinal tract—arguably the core machinery that keeps us going—together account for only 7 percent of our body weight, and yet burn a whopping 70 percent of our energy.) On top of that, the structure of the human body limits our total available energy reserves to roughly two hours of continuous, vigorous activity. At that point we start running a deficit. So as animals that walk around with only two hours of gas in the tank, our big brain is something of a luxury. We need to feed it, but we can't just run around endlessly to do so. In fact, even as ancient hunter-gatherers we'd already developed habits and responsibilities that required minimizing our pursuit of food. We had other tasks to accomplish, ideas to ponder. We still do. That means we have to economize the calories we burn.

And the brain is great at economizing. For one thing, although the brain doesn't seem to burn more actual calories with intellectually demanding, creative, "System 2" work, research shows that people report fatigue after long periods of involved mental labor. Complicated, careful thinking wears us out somehow. So perhaps System 1 costs us less, and that's why we rely on it so heavily.

Decades of scholarship support this idea that careful decision-making is more costly than following our instincts. In 1908, the psychologists Robert Yerkes and John Dodson tested the effects of stress on mice as they learned new skills. They gave one set of mice an easy visual choice between paths in a maze, and shocked them if they chose wrong. To others they gave a harder choice, along with the same shocks. They found that under stress, easy tasks were learned more quickly than they were under relaxed conditions. But stress made *difficult* tasks much more difficult.

The Yerkes-Dodson law has been refined in human studies over the last century to theorize that stress makes instinctive tasks easier, whereas it tends to harm our ability to bring attention, memory, and problem-solving skills—the System 2 skills—to bear on a problem. When the going gets tough, the tough think less. Our brains seem to be built to rip through the unconscious stuff under pressure, get it over with and get out of here, and to chew on the complicated stuff rarely and slowly, only when we can afford the time. Think of it this way: under urgent, stressful conditions, when the body assumes that a bunch of calories are presumably going to be needed for running, shouting a warning, or climbing a tree, instinct does the job. Under relaxed conditions, when fewer calories are required, our more "human" abilities get a shot at the problem. In order to afford our brain, we need strategies that budget our calories.

Now carry that logic over to the modern world, where, unfortunately, our lives are full of cognitive demands, and most of the time our brain cannot afford them.

This is why study after study, like the ones in the previous chapter, has shown that what we think of as well-considered choices are in fact offhand, instinctive decisions. We think we're rationally evaluating that used car, that rope swing over the lake, that plane ticket for a weekend trip. In fact, we're probably going on instinct much of the time. In 1992, the psychologists Nalini Ambady and Robert Rosenthal surveyed dozens of studies, especially a surge of behavioral research in the 1970s and 1980s, to arrive at the notion that humans are prone to "thin slice" judgments, based on quick glances at one another. An inappropriate outfit, a fistful of cash, and a missing shoe can generate a vast sequence of assumptions in our minds. "We communicate our interpersonal expectancies and biases through very subtle, almost imperceptible, nonverbal cues," they argue. "These cues are so subtle that they are neither encoded nor decoded at an intentional, conscious level of awareness." We choose a restaurant not by grilling the host out front about the sourcing of the ingredients or the training of the chef. It's a rare customer brave enough to walk in and ask a table full of diners how their meal is. We stand at a distance, squint at the awning, maybe take a quick look at the menu, and roll the dice.

And one reason we continue to make these unthinking choices is that they're easy to make and tend to be more or less useful. Ambady, whose work is described at length in Malcolm Gladwell's *Blink*, made the case in a 2010 paper that the thinnest slices of our behavior not only "provide information regarding personality, affect, and interpersonal relations," but our brains may not feel the need to go much deeper. Those thin slice judgments are often acceptably correct, and perhaps as a result we don't bother to really examine them. Why don't we take longer? Why don't we engage more mental machinery and examine more evidence? As Ambady put it, thin slice judgments are easy and effective. "They do not seem to drain cognitive resources and can be accurate even when processed in parallel with other tasks."[2]

But surely we're only making snap judgments about one another in contexts where the decision doesn't matter? When we interview someone for a job, or sit in judgment of a defendant on a jury, or review an applicant's qualifications for a loan, we must be engaging System 2, our higher cognitive functions, right?

In 2005, Princeton psychologist Alexander Todorov and three colleagues published the results of an experiment. They flashed pairs of faces past 1,000 subjects, giving them only a second in which to form an opinion of the two, and then asked which one seemed more competent. What the subjects didn't know was that the pairs were opponents in three years of House and Senate races. And when Todorov and his team analyzed the results, they discovered that the intuitive preference for one face over another lined up, statistically, with *which candidate had gone on to win*. Of course, the authors wrote, "from a rational perspective, information about the candidates should override any fleeting initial impressions. From an ideological perspective, party affiliation should sway such impressions." But unfortunately, "from a psychological perspective, rapid automatic inferences from the facial appearance of political candidates can influence processing of subsequent information about these candidates." And here's what was particularly alarming. In a nod to Ambady's findings that quick judgments can often be sound judgments, Todorov and his colleagues wrote that "although research has shown that inferences from thin slices of nonverbal behaviors can be surprisingly accurate, there is no good evidence that trait inferences from facial appearance are accurate."[3] In other words, snap judgments are often correct, so we port them over to modern tasks like evaluating the worthiness of a candidate from a glance at her face, even though science hasn't shown that that works at all.

Once upon a time, our highly efficient System 1 was a fabulous way of finding fruit and spotting danger, allowing us to prance with ease through tasks that make million-dollar robots look like drunk children. But now we're using this same rules-bound system to make complex choices for us in a world built on inaccurate assumptions about how far we've come. We're using a system built for survival to do things for which it is not adapted. We've built institutions, theories of governance, and expectations of one another that have absolutely nothing to do with the landscape in which we originally evolved our mental habits millions—or even tens of thousands—of years ago.

But those ancient decision-making systems are also full of ugliness. There are tendencies we evolved over hundreds of generations that served a practical purpose at the time but are totally unacceptable to our modern values. That's why we've worked so hard in the modern world to establish written values, abstract systems of governance and cooperation, and difficult but essential social norms, to pull ourselves up above those ancient tendencies. The trouble is that the ugliness is still inside our programming and can be easily accessed without our knowledge, even *while* we're building technological systems intended to get past it.

Chapter 4

CLUSTERS

WHAT ABOUT THE beauty of our natural state? Shouldn't we be striving to live as we did in ancient times, gathered peacefully around a fire, sharing our food? Considering that so much of our modern world seems to be disorienting the ancient systems we use to navigate it, shouldn't we be steering ourselves back toward the pure, honest, sustainable ways of ancient life?

"No no," says Mahzarin Banaji. "Our natural state is terrible, horrible. You'd never want to go back to that."

Banaji, a professor of social ethics in the Harvard psychology department, is an electric personality, crackling with humor behind thick, colorful glasses. But her work, studying the ancient biases we don't recognize in ourselves, is a bummer. She finds her own path to optimism, though, and often begins her speeches to large audiences with a spectacular bit of dark humor. "I'd like to congratulate you," she'll say, "on the sheer diversity of this crowd. You all come from wildly different backgrounds, you are drawn from different genders, ages, social status, races, religions. In the tens of thousands of years in human history, you are the first group this large and this mixed up in which no one is going to die before I'm finished speaking. There won't be a riot, no one will be burned for being different. It's amazing. Congratulations!"

In 1996, as a graduate student in psychology at the Ohio State University, Banaji was helping to design experiments about bias. At the time she thought very highly of herself as a fair-minded person. And if anyone's life story suggests modern conscientiousness, an immunity to ancient prejudices, it's Banaji's. She'd grown up Zoroastrian in a Parsi community in India, mentoring other students from the age of five in the small school her mother and aunt ran in their home before rocketing through a series of schools and universities and winning a fellowship to study in the United States. "I figured I was about as unbiased as one could be, and I thought very little of certain parts of society and *their* biases," she says.

And so she sat down one day to take an early version of the bias test she and her colleagues had designed, feeling perfectly ready to have her self-image reflected back at her. Up to that point, psychologists had largely assumed that they could measure bias by asking people to volunteer their own attitudes toward other people. But that sort of thing only revealed what people *thought* their biases might be. The test in front of Banaji was different. She was part of a group trying to design something that would draw out "implicit bias": unconscious attitudes, ones of which we're not only unaware, but we also might be horrified to discover. It's a common term today—presidential candidates have used it on the campaign trail, and it's the subject of countless corporate training programs—but at the time it was a revolutionary idea. The test asked Banaji to look at a series of faces, and as each appeared, press one of two keys to identify the face as Black or white. At the same time, the test asked her to use the same two keys to label occasional words as either positive or negative, one key assigned to each. The test had seven stages, during which it kept switching up which keys did what; one key would be used for both white faces and positive words, then in the next test it would be used for white faces and negative words. The other key would be used for Black faces and positive words, then Black faces and negative words. It's laborious to describe, and it's a dizzying test, just tedious enough to cause the subject to lose focus, and something about that, along with using the same finger to denote both a racial identification and an emotion, teases out long-held, instinctive connections we make between the two. In the end, the test told Banaji she had "a strong automatic preference for European Americans over African Americans."

"I had two reactions," says Banaji, now a towering figure in the study of bias. "The first was to assume that the test must be wrong. I mean, if my two fingers on a keyboard can't produce the result I'd expect about my own attitudes, the test must be flawed! And then, when I discovered that the test wasn't wrong, I was embarrassed! Mortified!"

Banaji and her colleagues Tony Greenwald and Brian Nosek put the test online in 1998, hoping that perhaps five hundred people might take it in the first year. Instead, forty-five thousand people took the test in the first month. That popularity raised the question of selection bias. They worried that only liberal people who consciously wanted to fight prejudice would take it, skewing the results. But that quickly faded as people of all backgrounds poured in, day after day. "I remember one day looking at it, and we noticed 400 people from Topeka, Kansas, had taken the test," Banaji says. "And we realized, 'Oh, it was a school that sent a large number of kids to us.'" Accounting firms, military units, museum staffs, all came through. "Neo-Nazi groups used to visit in the old days—we used to get messages from them." And the flow of data has not slowed. "For these 22 years, thousands of people come to it every day," Banaji says.

Taken by more than 30 million people since it first went live, the Implicit Association Test, or IAT, uses simple keyboard mechanics to measure our instinctive attitudes about gender, age, race, sexual orientation, disability, body weight. And because the test has more or less remained the same for so long, as new participants take it, the IAT at this point represents a comprehensive longitudinal survey of how attitudes have changed over the more than two decades since it began.

The good news is that certain attributes—like sexual orientation—are losing their stigma. "Perhaps it's that people often know someone who is gay, they can relate to them personally, they're a family member, a child," Banaji says. Senator Rob Portman of Ohio had been, for most of his career, staunchly opposed to legalizing same-sex marriage. But then his son Will came out to him and his wife. "And that launched an interesting process for me, of rethinking my position on the issue," he told CNN correspondent Dana Bash in a 2013 interview. "I now believe that people have the right to get married."

Banaji waves off Portman's change of heart. "I wouldn't extend any kudos to Senator Portman," she scoffs. "Apparently you need a personal experience to change!" That's not going to make any difference in other areas of bias, she says. "Your son is never going to come back after his first semester at Yale and say to you 'Dad, I'm Black.'"

And unfortunately, Banaji's data shows that racial bias, while it has fallen ever so slightly in the last decade, is still more or less as it was. Banaji cuts a straight horizontal line through the air with her hand. "Race is like this," she says. "I think people would be very surprised to discover that our racial biases really haven't changed at all."

But what about seminal cultural and historical moments? *The Cosby Show? Oprah?* The election of nonwhite candidates like Kamala Harris to the highest offices? Sadly, they don't move the needle. "The day after Obama was elected we were looking at the test results to see if anything moved," Banaji says. "And no." This isn't to say that our conscious attitudes can't be changed by a singular event. "Often people will report an individual experience with somebody of another group that changed their minds completely. This is why we encourage foreign travel. But I think for what we are talking about,"—long-term, large-scale, society-wide changes of attitude—"this bean-counting little machine"—your brain—"is saying 'x goes with y.'"

As we've seen, our brains use long-term patterns to take cognitive shortcuts. And Banaji explains that changing those patterns requires huge new amounts of counterfactual data. "What we're talking about here is something that operates at a level we're not conscious of," she says. "'What's the probability that when I stop at this newspaper stand in New York City that the owner will be South Asian?' And because the probability is so high, we will make that association. For it to go away, we would have to show them many, many, many instances of walking up to a newspaper stand, and having a quite different person be there."

Awful as it seems today, there is a reason for all this. She describes bias as a lifesaving adaptation to a brutal world, one in which we lived in small clusters, where any outsider might be a threat. Being able to instinctively and rapidly identify someone as an outsider was a wonderful thing. And being able to react appropriately based on assumptions about one's surroundings through repeat exposure to the color of ripe fruit or the movement of a predator in the firelight was a time-saving way of staying alive and protecting our communities. Researchers even believe there may be psychological mechanisms at work that help us avoid disease by avoiding outsiders, sticking to our own routines, and being generally socially conservative, a set of unconscious strategies known as the behavioral immune system.[1]

But today, our relentless statistical bean-counter minds are a terrible disadvantage to us when it comes to living according to our new, higher ambitions as human beings. Our lives have changed; our unconscious habits haven't. We aren't built to treat new people fairly, to inquire about their lives and invite them into ours. Millions of years of evolutionary pressure served to crush open-mindedness, because open-mindedness would have gotten us killed. Any shred of it we have is effectively made up. We put equitability on like a costume, a smooth modern aspiration draped over sharp, dangerous, ancient programming.

I once interviewed Lord John Alderdice, who grew up during the troubles in Belfast, and was part of the political movement that built the Good Friday Accord, the result of years of negotiations for peace in Northern Ireland. He's a member of the House of Lords and now studies conflict around the world. He's passionate about the work that wildly different people do to find common ground, but as a practicing psychiatrist and someone who has spent decades trying to facilitate that work, he believes it runs contrary to our natural instincts.

"The whole process of globalization, the freedom to trade and travel and move about, not just you going to other countries, but them coming to you—there's a certain percentage of the population, perhaps 15% or so, who like that idea," he says. "And these in the main are people who inhabit newspapers and academic institutions. They feel happy and comfortable about it all. But what they don't get is that most people don't feel that way. They just want to holiday and come home. What they don't want is other people coming, not for a holiday, but bringing with them their cultures, and doing so at a rate of change they can't cope with."

Consider just how new it is to be able to think the way we do. Our closest common ancestor with modern-day primates walked the earth perhaps 30 million years ago. We began to resemble our current form roughly two million years ago. But it was only perhaps

200,000 to 70,000 years ago that meat-eating and heaven knows what else pushed us to wonder what lies beyond our bodies and our immediate needs. The neocortex, as neuroscientists call it, is the part of the brain that gives us the ability, acquired very recently in the evolutionary sense, to think beyond ourselves—to not just pick up cues from others of our kind, but put ourselves in their shoes, stay up late worrying about our future and past as individuals and as a group, empathizing and negotiating and coordinating with each other on a constant and largely peaceful basis.

Our ancient selves developed the ability to quickly recognize, ostracize, and even punish difference. Those ancient humans could not have done what you do each day. And we're not just talking about the big stuff, the peace negotiations between warring peoples that experts like Lord Alderdice help to facilitate. We're talking about any of the trappings of your daily life. Ancient peoples could not have woken in a home respected as their private sanctuary and sat down to work all day with a wildly varied group of other humans. Someone would murder someone, as Banaji likes to point out. They certainly couldn't have voted peacefully for leaders they'd never met and will never meet, sued one another in court, agreed to treat certain pieces of paper as having value, or placed the smaller of their two carry-ons under the seat in front of them. The trappings of our lives are all astoundingly new and alien, and when we see the mismatch between our cognitive equipment and the work we do with it, and then realize the degree to which our ancient biases predetermine how we negotiate the corridors of modern life the way TN negotiated the hallways of Beatrice de Gelder's lab, it becomes clear that this balancing act is very fragile indeed.

Kahneman wrote in 2003 that his work on System 1 and System 2 drew from two broad hypotheses about human decision-making. "The first is that most behavior is intuitive, skilled, unproblematic, and successful." Automatic decisions, after all, have kept the species alive all this time. But the second hypothesis is where we get into trouble. "Behavior is likely to be anchored in intuitive impressions and intentions even when it is not completely dominated by them." You and I are more or less always influenced by these unconscious systems even when we've exercised *some* rationality.

Still, haven't we created ethical standards and governing principles and legitimized organizations that correct for those outdated instincts? Can't we reason our way out of this? I believe it's one of the miracles of modern life that we can agree as a society to stay in our lanes on a highway, not to mention that we can go to court peacefully when we've been wronged. But sadly those systems don't mean we've left our ancient instincts behind. Just as Alexander Todorov and his colleagues found that we vote with those instincts, even when we think we're voting with our higher functions, the Yale psychologist Yarrow Dunham has found that in-group/out-group biases are so ready to pop to the surface that he can draw them out of children as soon as he meets them.

Dunham is a friendly, open-hearted sort of guy, quick to laugh, warm with strangers. Not exactly the kind of person you'd expect to have spent years uncovering powerful discriminatory tendencies in children. His lab at Yale directly tests the ways human behavior is, as Kahneman wrote, "anchored in intuitive impressions." I watched an experiment in which he assigned children to arbitrary groups, and then had them share what they assumed about their own groups and others. It's terrifying to witness.

In his experiments, Dunham asks small children, between three and six, to spin an arrow on a wheel. The arrow lands on either green or orange, and Dunham presents the child with a T-shirt in that color, pronouncing them a member of the green or orange team. By that point, he tells me, "I can have them making outrageous assumptions about each other within 90 seconds."

I watched him do this several times. He'd begin by testing the children's assumptions about their own team and about the other. Which of these two kids, green or orange, is most likely to give you candy? Which of these two would you want to be friends with? The results were immediate, and consistent, and awful. He showed one girl, dressed proudly in her new orange T-shirt, an image of an ambiguous scene. In it, a child in orange lies sprawled beneath a swing set while another child in green stands behind the swing, looking concerned. "He pushed her off the swing," the girl told him. But later, when a boy in a green shirt examined the same scene, he told Dunham that the girl in orange must have fallen, and that the boy in green was about to help her up. Same scene, same facts, interpreted opposite ways.

"They start out with just a small bias in favor of one group over the other," Dunham explains. But over time, as he shows them more of these scenes, their biased interpretation of the scenes begins to become a body of evidence in their minds. And that's when assumptions really start to lock in. "When I later ask them about the group, they have a little stockpile of what they think are facts to draw on. 'Oh, well, members of that group are just mean, they're pushing people over all the time, they're stealing people's money.' So now they have essentially evidence in favor of their bias, and now you can imagine this leading them even more to think that that initial bias is justified."

Now imagine playing those assumptions out over a life span, Dunham says. "Over time the accumulated weight of all of that is a very entrenched view of your own group as positive."

Dunham's work is part of a decades-long effort to find out why we so quickly give ourselves over to biases about even the most arbitrary and petty of groupings. Turns out we're just built for it, somehow. In the 1950s, the social psychologist Henri Tajfel, a Polish Jew who survived the war concealing his background in a series of German

POW camps, sought to find out why even ordinary Germans had gone along with Nazism and its horrors. By 1970, he was conducting experiments at the University of Bristol in England on the notion of a "minimal group paradigm." The idea was that it took nothing more than membership in a group—people connected by anything, from similar Christmas sweaters to nicknames—for the members to begin favoring one another over outsiders. Ever waved at a passing driver because they happen to be in the same kind of car you are? That's the idea here.

It turns out Tajfel wasn't immune to the effects he studied. A pair of researchers put together an investigation in 2019[2] that found Tajfel was not only a serial sexual harasser of the women in his lab, but remained "resolutely uninterested" in applying his findings about in-group/out-group preferences to gender, and neither treated women as intellectual equals nor pulled them into meaningful collaborations. In Europe, the organizers of a prominent academic award named for him recently announced they'd be renaming the prize.

But his findings about a minimal group paradigm are as strong today as his unfortunate personal demonstration of it. Hundreds of experiments have reproduced the same dynamic. Slap us into random groups and we'll assume ours is superior to another. (As the psychologists Jeffrey Lees and Mina Cikara have found, perceptions of what the "other side" believes are also often inaccurately pessimistic, even on subjects where both sides largely agree on the issues.) And as Dunham explains, that tendency camouflages itself in our minds as objective judgment.

HOW DO WE behave when these tendencies are camouflaged, when we think we're making our own choices, but we're actually being steered by ancient instincts? Frankly, this sort of thing is everywhere in our lives, but the ways it takes some of us to the very edge of acceptable society are especially revealing. I've met many brave, resourceful researchers in my time reporting on technology and human behavior, but the researchers at Artis International have perhaps the most daunting job description of them all: find people so committed to a cause that they'd fight and die for it, then get close enough to them to truly understand their commitment. The researchers that make up this think tank seek out, befriend, and study the world's most devoted fighters in the bloodiest conflicts on earth. And in the process, they're discovering that the systems that make us human may not make us the kind of humans we'd like to think we are.

Lydia Wilson, an Artis research fellow, studied history at Cambridge and speaks Arabic. She has lived and worked in Syria, Beirut, Kosovo, and Jordan, and is considered one of the more fearless members of the team. Often her interviews take place where even journalists don't go. "At one point I was interviewing a Peshmerga general [part of the military forces of Iraqi Kurdistan] on the front line," she remembers. "As we were driving away afterward, he asked 'didn't you hear the ISIS attack during your conversation?' I hadn't. In the transcript of the interview I ask 'Is it raining?'"

She laughs. "The Peshmerga are pretty extreme, and they always talk about Kurdishness. They won't blow themselves up, for instance. It's not Kurdish. But hospitality is a big thing for them. Very Kurdish. So it turns out they'd ordered their troops not to fire back, in case the shooting scared me."

In a 2017 study published in *Nature Human Behaviour*, Wilson, Artis cofounder Scott Atran, and several colleagues detail the attributes of "devoted actors," people like the Peshmerga fighters, defined by powerful and frightening characteristics: "commitment to non-negotiable sacred values and the groups that the actors are wholly fused with" as well as "readiness to forsake kin for those values."[3] They're banding together using ancient, autonomous systems that even our primate cousins display, but they've layered in some new stuff—a ready-to-die commitment to certain things—that's uniquely human.

The devoted-actor paper is based on field interviews with Peshmerga, Iraqi army troops, Sunni Arab militiamen, and captured ISIS fighters. In conversations with the researchers, these men offered spontaneous explanations for their actions that echoed patterns Artis researchers had encountered before. (The study was supplemented by online surveys of more than six thousand noncombatants in Europe.) They demonstrated an astounding loyalty to abstract ideals. They also demonstrated greater loyalty to their fighting group than to their own families. And when offered the possibility of trading the battlefield for earthly comforts—money, a safe life in a peaceful nation, a better life for their children elsewhere—they often became blisteringly angry.

What makes these fighters so devoted? The authors tried to determine the factors that directly compel everyday human beings to split off from the simple goals and comforts of family and community and voluntarily throw themselves into warfare. By the end of the study, they identified a few "sacred values" powerful enough to do it. And in studying those values, they may have identified dynamics that compel you and me, as well.

The study defines a "sacred" value as one that is impervious to temptation, utterly nonnegotiable. "To measure sacredness, we investigated willingness to trade-off values in exchange for material benefits, whether for individual or collective gain," the authors write. "Absolute refusal to contemplate such trade-offs was taken as an indicator of a sacred value."

What were those values? For Peshmerga and Iraqi Kurdish troops, they included a commitment to an independent Kurdistan, and to the dignity of "Kurdeity." For Sunni Arab militiamen, they included questions of Sharia law and of "Arabness."

And the groups had a powerful confidence in their own spiritual strength, rather than judging their prowess on the battlefield through a rational military assessment. When shown diagrams of certain amounts of muscularity and asked about the strength of their bodies and their comrades' bodies, they had an outsized opinion of their group's physical power. Shown the same diagram and asked the same questions about American troops, they tended to consider themselves far more physically powerful. "They argued that most important was spiritual formidability," the researchers wrote.

In the end, the Artis researchers came up with a list of the necessary ingredients to make a "devoted actor." It requires that a person subsume their individual identity within a group identity, usually prizing that group even over their own families. They possess abstract beliefs about freedom and dignity that cannot be swayed by material enticements. They evaluate themselves and their foes based on their spiritual strength, rather than on their numbers or their weaponry. Scott Atran told me that he believes there's some evolutionary purpose to all of this.

"These transcendental conceptions that go beyond individuals make the group immune to regular scrutiny, and that turns out paradoxically to give us incredible group strength and personal strength," he says. "That's what got us out of the caves."

"You need ways to keep these bigger and bigger groups together. These systems of strange ideas—transcendental ideas that can't be empirically tested—become ever more important. For one thing, you have to ensure that the members of your group are blind to exit strategies. So no matter what enticements you offer them, they reject them. 'Better dead than red,' that kind of notion."

Nafees Hamid, a research fellow for Artis, has worked for years to understand newly minted fighters for ISIS, the banned UK jihadist group Al-Muhajiroun, and others. His research has taught him a few things about what sorts of values set off deep, automatic, emotional reactions, and which values get a calmer, more considered reaction from the brain. For a 2019 paper he, Atran, and others interviewed Al Qaeda sympathizers in Europe, drawn from specific immigrant communities, and even imaged the brains of those that could be convinced to spend time in an MRI machine. Hamid says the findings suggest that sacred values may be processed by the part of the brain that learns and obeys concrete rules: the stove is hot, the cliff is dangerous, et cetera. "Truly sacred values seem to set off that deontological system," Hamid says. "And when it's non-sacred values, it's the executive control portions of the brain," he says, the portion that slows down, thinks things through, and processes our higher thoughts. "Basically, the reaction time is slower for the non-sacred values."

Topics that trigger our strongest feelings don't take advantage of our mental horsepower. In a sense, tribalism and religion and the rest of it use a separate processing system. As Hamid, Atran, and their coauthors wrote, "Sacred value choices involved less activation of brain regions previously associated with cognitive control and cost-benefit calculations." If that's how the brain works, they suggested, "this would involve different degrees of cognitive effort in both situations: sacred values would work as a heuristic making decisions easy to solve, or cached-offline, whereas decisions regarding non-sacred values would involve some degree of calculation."[4]

So what actions violate sacred values and set off instinctive reactions? "One of them was drawing a caricature of prophet Muhammad," says Hamid. "Another is Sharia law in non-Muslim lands. Stuff about Kashmir and Palestine. That stuff is the low-hanging fruit of sacred values." The non-sacred values, which provoke a more measured response, "were US military occupation in Muslim lands, drone strikes, whether there should be Islamic teaching or sacred food in their kids' schools."

In effect, Hamid says, the stuff that actually affects the participants' day-to-day lives doesn't fire up the most powerful decision-making systems in their brains. "Sacred values are more broad, abstract, transcendent," he says. "ISIS isn't spending time talking about halal food in public centers."

It seems that it isn't Maslow's hierarchy of needs, or any other simple taxonomy, that governs these devoted actors. Abstract values—values that arguably make no material difference in our lives but speak to who we want to be—are more easily elevated to a sacred role.

The difficulty comes when one group is devoted to sacred values that conflict with the sacred values of another group. And that difficulty, Hamid says, makes him pessimistic about helping people get along. "Turning tribes into cosmopolitan cities, breaking down borders, seeing us all as humans—that takes a lot of trust and confidence and not feeling marginalized," he says. "The way that our brains have evolved is to be much more tribalistic than that. Our political structure goes against the structure of our brains."

What strikes me about the Artis team's findings is that they represent one of those evolutionary detours that began on the central path—tribe, transcendental commitment to our groups, social cohesion above all else—but veered off toward political disagreement and territory and irresistible conflict. And now that we're speeding along, generations later, under the power of our ancient in-group/out-group tendencies, we're adding the accelerants of media and technology and capitalism.

IN THE MODERN landscape of information, in which companies make money the longer we remain on their platforms, and in which they hand us content that's been curated by recommendation algorithms and dressed up by "user experience" designers, our ancient instincts are being strummed for maximum resonance. Experts have talked for years about the radicalizing effects of social media, the way that it can amplify our grouping tendencies, but I've always naively assumed that to be something that happens only to people without the necessary presence of mind to see things clearly. I realize now that I was deeply misguided, that our ancient instincts are being accessed at the same time that we're being convinced by the technology we interact with that that's *not* happening. Mak Kapetanovic taught me that.

Kapetanovic was twenty-two when we met. He grew up in Jacksonville, Florida, the son of immigrants from Bosnia. His mother died suddenly of a stroke when he was sixteen, and an already socially isolated life became even lonelier. He looked online for community and kinship. He was already playing a lot of video games, he says, and had encountered what he describes as a lot of casual racism in the chat and banter there, a sort of swirling vapor of white supremacy that was unmistakable but hard to pin down. And after his mother's death, he began spending hours on online forums like 4chan, where he found, in memes and pictures and half-baked statistics, the beginnings of a new ideology and personal identity.

He's a slight, friendly, soft-spoken person. He plays in a band. He was working at a restaurant and planning a return to college when we spoke. And he described his lapse into what eventually became an enduring obsession with white nationalism as a slide so gradual he lost track of its progress. "It's a really, really slow sort of descent, like a very slow escalator," he told me. "You can't see the floors on the escalator."

But looking back, Kapetanovic says he wound up in a very different place from where he began. His parents were immigrants from Bosnia, Muslims escaping the persecution taking place there. But now their son was adopting views on race, immigration, and Muslims that he has a hard time explaining today.

"I come from a Muslim family," he says. "I come from an immigrant family. I come from a place where during the war, you know, people like me that were ethnically Muslim were ethnically cleansed. Genocide. And for someone like me to then come to America and, you know, be born here and grow up here, and then, you know, be okay with the kind of ideas that I—that I was okay with, I mean, that's surreal."

He says that when he began exploring online forums, he had the conscious goal of *fighting* against racist ideology. "One of the first reasons why I started browsing the political board on 4chan was to look at and laugh at racist people," he says. "But just being in that environment, being around so many of those people and just reading what they have to say about things . . ." He pauses. "It can change your mindset in ways that you don't realize." He also says that YouTube, where he'd go to investigate terms like "race realism," consistently funneled him toward increasingly extreme content about crime, race, and immigration. The story he tells is a story I've heard over and over again: people thinking they're accessing the higher functions of their minds by investigating things online, only to find that over time, they've wound up empowering their most primitive instincts, to disastrous effect.

The identity he found online was everything. "Most people would look at me and say I'm a white person," he tells me. He's fair-skinned, with light-brown hair. "But I'm also a first-gen immigrant. It was really hard for me to find some sort of identity to latch on to. I didn't really feel American enough. I didn't really feel Bosnian enough." And so when his nights on YouTube and his hours-long chats on 4chan and various games helped him find a community, he grabbed on. "When I saw them—sort of rallying around—behind, you know, this sort of ideological pull—I thought, 'Well, okay, I guess I'm over here too.'" As Dunham could have told him, it feels really good to put on the orange or green T-shirt. And as Banaji could have told him, his conscious mind was not in control. In the end he says he wound up believing in the sort of racist ideology that had once killed portions of his family in Bosnia and drove his parents to flee the country.

When we met, Kapetanovic had abandoned all of that, and he planned to study anthropology in order to fight extremism. His turning point was the 2019 murder of fifty-one people at a mosque in Christchurch, New Zealand, when he realized that the shooter, Brenton Tarrant, whose online influences were on display to the world after his arrest, had been circulating the same memes Kapetanovic had. "A lot of the same ideas were fed to both of us, but he killed 50 people, and now I'm fighting against that ideology," he said.

I asked him whether he could imagine an alternative outcome in which he might have gone as far as actually hurting someone the way Tarrant had. He spoke quietly. "It's really scary to think about. I'd like to think no, never." He paused. "Maybe if my father *and* my mother passed away, maybe that would've done it. Maybe if I didn't have those few close friends that I did, maybe that would've done it."

Ultimately, he says, he believes technology is what accelerated his descent on the escalator, while preventing him from realizing it. And he worries that the next kid won't pull himself back, because, he says, humanity simply isn't ready to counteract that sort of effect. "We as a society, we—we've never really had to deal with ideas spreading in this way," he says.

THINK ABOUT IT this way. Artis has found that devoted actors fuse their individual identities with those of groups. This tendency hands a powerful tool not just to extremists, but to any group—gun-rights activists, democratic pollsters, pug rescuers—that wants to recruit members, raise money, or influence policy. The innocent act of joining a Facebook group that draws us in among the like-minded may in fact be acting more powerfully on us than we're aware. (This may help explain why Mark Zuckerberg made groups a strategic focus of the company in 2017: whether it's skateboarding or an ideological position, the ties that bind a tribe make membership very compelling.) Our brains are built to have a much deeper emotional reaction to who we imagine we are and what we can do with other people like us (that's the power of notions like freedom, country, principle) than to who we actually are (parents, friends, drivers of children to school). Mak Kapetanovic found that out firsthand, although it took him years to realize it had happened. And as Hamid has found, we're more willing to participate when we don't engage the parts of our brains that rationally weigh costs and benefits. What company wouldn't want to sell to that part of our minds?

While we like to imagine that the tribalism and deep emotional reactions to sacred, abstract values of devoted actors are a long way removed from who we are, that may not be the case, at least not for much longer. Because more and more, it's going to be our dominant form of social interaction, and more and more, companies are learning to make money from it.

At one time, this sort of unconscious decision-making and group formation, and the way that technology companies catered to it, was mostly considered a social aberration, an innocent addiction, a time waster. Our video games, chat rooms, and niche pornography were just the profitable eccentricities of a newly connected era. But now it's clear that what Artis is measuring in conflict zones is taking form here in the United States as we form online groups committed to conspiracy theories like Pizzagate and QAnon. Sacred values and group identity cause us to abandon our critical faculties. In Hamid's experiments, MRI scans of people angry about a cartoon of the prophet Muhammad or about the occupation of Palestine showed that their cost-benefit centers weren't lighting up. That would clearly have been true of anyone inside the Capitol building on January 6, 2021. What better evidence that sacred values turn off our better judgment than people storming up the steps without a rational plan for what they'd do when they got inside, and who hadn't considered that doing so, without hiding their faces, while livestreaming themselves on YouTube, meant felony charges awaited them on the far end?

I spent January 6 monitoring the online personalities livestreaming the events at the Capitol. I watched commentators narrating what was happening, putting up simultaneous feeds from the steps, the doorways, the interior of the building, on YouTube, Twitch, Discord, and other platforms. And this wasn't just ideological. They asked for money as they did it. On YouTube, so-called Super Chat windows were a roaring river of comments, with the occasional commenters paying $20, $50, or $100 to have their message briefly pinned to the top of the discussion.

A few days later I spoke to Hamid and asked him whether what he'd seen on January 6 reminded him of the jihadist behavior he'd studied. He told me it did. And he told me that the online communities that had helped to coordinate the events, and then celebrated when they took place, had facilitated the most powerful currency of the devoted actor. "The most devoted people are people who are completely ensconced," he said. "That is their only reference group. Oftentimes that is their only community, they've totally identified with them. And so for them, they're looking for a social reward from their fellow jihadists, white nationalists, conspiracists."

THE GROUPING AND networking of human beings has created modern outcomes like these that humans couldn't have imagined even a few hundred years ago, and that our fundamental, ancient programming has not prepared us to resist. As organisms equipped to detect snakes and fire and close relatives, we now live in a world full of impossible abstractions. We vote to elect strangers, we are policed by appointed members of society according to imaginary laws, we write out long, entirely abstract contracts by which we form and dissolve personal and professional partnerships. All of this stuff—all of it—is invented. And more important, it plays on the unconscious systems of our minds while simultaneously convincing us that we're using our higher, more recent consciousness to move through the world. The modern world is mostly System 1, dressed up as System 2.

In considering how technology will shape us in the coming generations, we are confronting the growing realization that the ideals we hold around fairness and freedom and rationality have not been achieved, perhaps because we are not fundamentally built, as humans, to achieve them. And yet because we want to believe that a society built on our fundamental attributes is the best possible form of humanity, the technologies and systems that we've built to shape and organize human choice may have not taken into account that we're all winging it out here. And in fact it may be that the best version of living together involves mostly fighting *against* our essential nature—our biases against those outside our group, our tendency to believe nonsensical things if it wins us the approval of our tribe, our instinct to sacrifice practical considerations for unachievable principles—even as we build technology that uses our nature against one another.

Just by being human beings, we're caught in the first loop: a cycle of unconscious decisions and influences that keep us in a more or less predictable orbit. And as modern humans, we've only just begun to recognize that loop. Yet at the same time, we've built a second loop for ourselves around the first: an industry of manipulation and persuasion that draws on our unconscious tendencies to lock us into a circular flight path of consumption and acquiescence. Some of what I'm about to describe will look familiar. It might even look appealing. But keep in mind that it's a result of both our cognitive vulnerabilities and the urge to make money off them. Seen that way, the world we float through, shepherded from place to place, is both very new, in terms of the history of our species, and very alarming.

Chapter 5

GUIDANCE SYSTEMS

WHEN WE WERE kids, my parents took me and my sister to Disneyland in California a few times, usually as an adjunct to a visit with my grandparents, and I remember each occasion vividly. I loved the obscure narratives, and being jerked this way and that by the overlooked Mr. Toad's Wild Ride, and while I would never have admitted it then, I found a real bliss in the placid predictability and intricate miniature landscapes of It's a Small World. But my favorite experience by far, the one that stuck with me in my dreams and my drawings when we returned home to Connecticut, was Autopia, a winding set of roadways where anyone over thirty-two inches tall was allowed to *drive a car*.

I remember the volcano of anticipation as we waited in line, and when I got the signal I sprinted out onto the track to choose among the various beach-buggy-like numbers, their early-1980s bright green and purple and yellow metal-flake paint jobs glittering in the Anaheim sun. We'd all climb in and I'd grip the steering wheel, anticipating power and freedom as the attendant fired up the incredibly loud and undoubtedly terribly polluting little car. And I can remember in my body, as I write this, the half-second delay between pressing down on the accelerator and feeling and hearing the two-stroke engine roar to life and pull my family and their nine-year-old driver (me!) wherever I decided to go.

But of course I wasn't actually deciding anything. I always misjudged the first corner—it turns out there's a reason we don't let kids drive—and subjected everyone on board to the horrific thump of striking the hard guidance spine that ran down the center of the roadway, a sort of solitary curb that coursed the length of the track and violently forced the front wheels left or right if the person at the wheel didn't keep the car centered in the lane. Because of the spine, I wasn't driving at all, in the strict sense of the word. I was, in effect, only steering, and even that part wasn't necessary, strictly speaking. If I'd taken my hands off the wheel, that guidance spine would have just forced the car in the proper direction. My efforts had more to do with trying to keep the front wheels parallel to the spine to avoid the thumped reminder of its existence than with actually choosing where to go. But all of that didn't matter. I *felt* like I was driving. I felt powerful. I felt free. And that experience hit my brain so hard it created for me a lifetime's fascination with cars and planes and boats and the mechanized pursuit of speed and freedom, even though I was really only in charge of the accelerator.

Over and over again, we see that our behavior, which feels to us like free will and clear choices, is actually the result of guidance systems we're helpless to obey, whether it's our own brains or something external like a guidance spine. And yet the factors I've tried to describe in the earlier chapters allow our brains not only to make us believe we're making our own choices, but also to fill in a narrative of power and independence in the same way that Autopia infused me with the feeling that I was on the open road.

Let's talk first about what I view as the largest obstacle standing in the way of our seeing this dynamic clearly: Not only don't we have the control of ourselves we think we do, for some reason our brains also make us *resentful* of *other people* wriggling in the grip of forces beyond their control. It's an ongoing reason we're so easily manipulated by each other and the systems we build. Maya Bar-Hillel, who studied alongside Amos Tversky, did pioneering work on inaccuracies in human reasoning, and is now a professor emeritus of psychology at the Hebrew University of Jerusalem, asked me, "Why is it that when we encounter *perceptual* illusions, we get all smiling and excited?" Fun-house mirrors, hollow heads—all of that is somehow enjoyable. "But when we hear about cognitive tendencies, we get all tense and defensive!"

"All these things that we do," she told me, "it's not because we're stupid. It's because we're human."

Sendhil Mullainathan, a professor of computational science and behavioral science at the University of Chicago, who has worked on everything from behavioral psychology to putting artificial intelligence toward a whole-body understanding of cancer, has done groundbreaking work on the cognitive burden of poverty. And not only has he revealed that poverty is vastly more crippling than we knew, he's also shown that we unfairly misjudge people suffering under its weight.

Mullainathan was drawn to the subject in part by reading about the infamous Minnesota Starvation Study, conducted in 1944, at the height of America's involvement in World War II. Huge swaths of the world were starving at that time, but little formal research had ever been done on the biological and physical effects of going without enough food. Researchers selected thirty-six males from a group of more than two hundred volunteers, many of them conscientious objectors who had agreed to serve the war effort but wouldn't engage in violence. They were chosen for their good physical and mental health, as well as their collegiality under pressure, because they were about to be placed under tremendous amounts of it.

The plan was that every man—they were all men—would lose 25 percent of his body weight. For three months researchers fed them a standard diet, 3,200 calories a day, mostly potatoes, pasta, bread, and other foods widely available in Europe. Then they suffered through six months at only 1,570 calories per day. The before and after photos are a nightmare: men with the already-lean physiques of a more constrained era transformed into walking skeletons. And they weren't allowed to just sit and suffer: they walked twenty-two miles a week, performed physical and written work, and were constantly interviewed. The negative consequences were clear: their resting heart rates slowed, their sex drives evaporated, and they reported irritability, depression, antipathy.

Something subtler leapt out at Mullainathan, though: starvation also transformed their unconscious lives. The researchers noted, in an offhand footnote, that many of the men took to reading cookbooks during the study—pretty uncommon for your average man in 1940, and a torturous choice for men consuming so little food. And a third of these men—drawn from all professions and walks of life—reported that when the study ended, they planned to open a restaurant. When the study ended, and the men resumed eating a healthy diet, that ambition evaporated. No one followed through.

Mullainathan went on to determine that poverty similarly reprograms the human mind. He and his colleagues spent five years investigating how poverty and hunger can affect our mental abilities as well as our bodies. And they found that just as being hungry makes people preoccupied with food, being poor will make them preoccupied with money. He and his researchers asked hundreds of people outside of grocery stores how much they'd paid for the items in their bags. People with enough money to easily purchase groceries admitted they couldn't remember. People who could barely afford what they needed turned out to know exactly what a tube of toothpaste cost. It shows, he says, that traditional assumptions about persistent poverty being a function of laziness, or disorganization, or some other personal failing are wrong. And it shows that someone experiencing poverty is intensely distracted at all times. They are literally impaired by it. He went on to write a book about this work with psychologist Eldar Shafir, *Scarcity: Why Having Too Little Means So Much*.

Other researchers have discovered similar hard, guiding dynamics at work in other parts of our lives. A study by Anuj Shah, an associate professor of behavioral science, also at the Booth School, found that when two groups of people—those with money and those without—were given the same instructions from a doctor for treating an illness, those with money tended to recall the instructions clearly. Those without money recalled only that the treatment called for medication and what the pills would cost. Remember Pötzl's work with Obszut, and everything that came after it? Our brains assemble a secondhand version of reality from the raw feed? Well, the research also suggests that people with and without enough material resources are experiencing distinct realities—literally seeing the world differently. Our brains are disabled by our unmet needs, and yet political rhetoric, self-help authors, and the plots of countless popular movies try to convince us that not only can we muscle our way to an independent life, but also that any claim to the contrary is a betrayal of us all.

"You would never ask the starved guy in the Minnesota study to lift a heavy weight, right?" Mullainathan asks me. "And you'd never blame him for not being able to do it. So why do we blame people who are poor for having a hard time doing certain things?"

All of this makes it that much more miraculous that we, as a society, *have* been able to recognize the power certain invented systems hold over our behavior. Just as, according to Banaji, the human brain needs huge numbers of counterfactual examples to stop associating certain qualities with certain races or genders, our society requires enormous amounts of evidence to begin blaming our behavior on a guidance system. It tends to happen only once the science has clearly established the way those systems interact with our biology to grab hold of us, so clearly that it can be proved in court.

AT THE AGE of fourteen, Sean David imagined becoming a doctor. Growing up in the 1970s, watching medical dramas after school, "I wanted to be Marcus Welby MD," he remembers. And so he took the first hospital job he could find after graduating from college. He became an orderly, scrubbing out surgical theaters after each procedure.

"You go in, you pick up all the biological waste and dispose of it. You put this sudsy soap on the floor and mop and hose it down, go room to room," he remembers. "The orthopedic cases were really rough, they use hammers and saws, they leave a lot of stuff behind." But he was building his mental capacity for observing, up close, the human body in distress, and the work was satisfying, in its way.

Then, in 1991, his supervisor found him as he was wheeling a bucket between rooms and took him aside. "Your father is in the catheter wing," he told David.

Cigarettes had always been part of David's memories of his father. "We'd go fishing and he'd be smoking cigarettes. I remember him putting bait on the line with a cigarette in his mouth. It wasn't unusual in those days." But now his dad was lying, ghostly pale, weak, frightened, in this hospital. He'd come in with chest pains. It was the first time David had ever seen him this way.

The cardiac surgeon told David and his mother that David's father had a heart lesion, likely from his decades of smoking, that would require a bypass. And as the surgeon walked back into surgery to get started, David realized "I had just finished cleaning *that room*."

Luckily, David's father survived the procedure. And the experience taught him what he wanted to do with his medical training. "It became clear that smoking is life and death," he says. "So many people don't survive that first heart attack. So emotionally, I'm very driven by this." After medical school, David began a research career studying why it is that patients who show every outward sign of wanting to quit smoking simply cannot do it.

Ultimately, he says he believes there are a few basic facts of cigarette addiction that anyone hoping to fight it off must understand. The first is that it plays perfectly on our cognitive shortcomings. First of all, when we are first offered a cigarette outside a party or after a concert, we are terrible at understanding the odds. "People consistently underestimate the risk of smoking," he says. "It's a 50% chance of premature death. That's one in two. A one-in-six chance of lung cancer. The risks are huge, but somehow risk has become attractive." Human beings simply lack whatever it is we need to truly internalize the danger, he says, and to avoid the deadly gamble altogether.

So people need to be sold on avoiding tobacco and fighting off the grip of addiction, just as the cigarettes themselves are sold. "There has to be a key promise, something to get you out of smoking," he argues. "We need formative research that finds out what motivates different audiences, because I haven't seen a sustained public health campaign that resonates with every segment of the population. Maybe if we could talk about lung cancer—but the trouble is, there aren't a lot of survivors, so there isn't a good advocacy group out there."

And there is no natural mechanism that will simply cause humanity to stop smoking. "People live to reproductive age while smoking, so evolution can't touch it, can't wipe it out," he says. In creating cigarettes—a consumer technology that delivers a consistently satisfying experience in a portable package—we've invented a deadly vice that is immune to natural selection.

This is the difficulty of the modern world. We invent technologies upon which we build new businesses and, eventually, entire industries. And yet those technologies and the money we make from them usually outpace our understanding of their risks and rewards. They often prey on our psychological frailties, excite us when they should revolt us, and, cloaked in marketing and social acceptability, evade our natural ability to recognize that something is bad for us and should be avoided. We go ahead and form businesses without a clear understanding of just how completely they may control the customers they serve, and how that manipulation may affect everyone else. In the case of cigarettes, which first entered industrialized mass production in 1845 under a French state tobacco monopoly, it was another 119 years before the US surgeon general Luther Terry issued a report that showed tobacco causes lung cancer, and another 55 years (2020!) before the US government finally required a national minimum age of twenty-one to purchase tobacco products. And all that time, secondhand smoke, which, it turns out, can slowly poison people physically separated from the smoker (in an adjacent apartment, for example), was affecting all of us, even if we managed to avoid picking up cigarettes ourselves.

We live in what Thomas Friedman calls the Age of Acceleration, when breakthroughs in fields like mathematics and computation are arriving every day, and entrepreneurs are just as quickly turning those breakthroughs into products. But this acceleration is dangerous. Like cigarettes, even the most manipulative and harmful systems we build are typically allowed to run free for generations before we sort out whether or not they're hurting us.

The behavioral science that's being outpaced is what we discussed in the first chapters of this book. And that science has really only just begun. Think about it this way: We've barely arrived at the kind of obvious, in-your-face data that tells us *smoking* is bad. So we are only at the very beginning of revealing all sorts of uncomfortable truths about common mistakes of judgment and estimation, our allergy to uncertainty, our desperate desire for assurance, our amnesia about the lessons of the past, the way we outsource important decisions to our emotions. And further complicating things is that these are, in fact, cognitive gifts, part of an evolutionary inheritance that has worked wonderfully at keeping us alive for the vast majority of human history. But we're now engineering products to appeal to those inner workings without fully understanding just how susceptible we are to being guided and manipulated. And while I believe it's clear that the mental and physical health of entire generations could be at stake, I also believe that capitalism, culture, and our conviction that we are in charge of our own destinies are all blinding us to the threat.

Today's entrepreneurs, including the ones who believe in the power of enlightened businesses to change the world for the better, seem to understand that our unconscious tendencies make us compliant users and customers. But they don't seem to recognize (or, perhaps, care) that we have almost no insight into the long-term effects of playing on those tendencies, even in the service of a noble goal. We're just barely able to spot the obvious, pernicious, short-term effects, such as the role of something like YouTube's recommendation algorithms in helping to radicalize a lonely teenager like Mak Kapetanovic. But we simply don't have the data or the accepted methodology yet to measure possible negative effects across nations and generations. And at the same time, we're deeply resistant to admitting when we're being manipulated: whether we just can't see it, we're too busy enjoying the product, or we're blinded by our resentment of those we consider somehow weaker than we are. (I think of the men starved by researchers in Minnesota returning to construction work after the study ended and trying to explain to a foreman that they won't be able to lift what they used to.) Meanwhile, companies continue to rapidly refine their understanding of our behavior and how they can influence it.

Perhaps it's encouraging that cigarettes, one of the first modern examples of a product that openly plays on our unconscious behavioral systems, are slowly being regulated, and that researchers are closing in on an understanding of their effects. But while the carcinogenic effects of cigarettes are clearly understood, the effects on our *behavior* are still not. Sean David—now a physician and researcher at Stanford—is working with psychiatrist and addiction specialist Keith Humphreys on creating new ways to measure our dependence on cigarettes. They're looking for biological indications of addiction and recovery. They've already found a few potential genetic indications that may help addiction programs tailor their treatments for particular patients.

This is good work made possible by the latest technology. And we need it. Because even now, when medical consensus has finally, clearly determined the health threat posed by cigarettes, medical science doesn't even have a clear way of measuring the extent to which someone is addicted, or their capacity to quit. We made cigarettes before we had any sense of how deadly they'd be. We certainly didn't know how we'd measure whether people were addicted to them. Billions of dollars were made in that industry before those questions even came up.

Until 2016, I was the kind of person who preferred to blame these sorts of universal human vulnerabilities on some individual failure, much in the way we often blame those in poverty for their situation. That year, I stood on a street corner in San Francisco, listening to a man in his early twenties named Patrick tell me about his heroin overdose. (He didn't give me his last name.) He had a faint wisp of a mustache, and if it weren't for his gray skin and sunken cheeks, I'd have mistaken him for a software engineer or the bass player in a band.

We met in the Tenderloin neighborhood, where the city's heroin and methamphetamine problem holds its most visible market. That summer, a deluge of fentanyl, a fast-acting opioid a hundred times stronger than morphine, had swept through the city. It caused nearly one thousand overdose deaths in just three months.

I was there with a camera crew filming a report about an overdose antidote drug called Narcan, and it was a desperate, stressful environment. People kept peeking through the windows of our car, evaluating its contents, even though it was parked directly next to us. We had to time our questions so that Patrick responded during the breaks in shouting, when disoriented passersby weren't arguing in the background. The manager of a harm-reduction program watched over us, and that seemed to be what gave us the occasional stretches of relative silence we needed to complete the interview. The whole scene was tense. And yet when Patrick began speaking, he did so easily and honestly, without any embarrassment or defensiveness, even as he talked about taking a drug that nearly killed him.

"About six months ago I was in Minneapolis living with my girlfriend," he told me. "I acquired some heroin out there that was a lot stronger than what I was used to in San Francisco." He described this the way I might describe being caught in the rain. "So when I used it, I OD'd."

We paused to wait out a shouting match down the street. Then he turned back to me.

"I was out completely. I didn't have any time to prepare for it or try to save myself or do anything. And my girlfriend luckily had trained in how to use Narcan." He shook his head gratefully. "She revived me in about five or ten minutes."

This was the first of roughly a dozen conversations I went on to have with people addicted to heroin over the next few weeks, but that first conversation with Patrick thoroughly dented my instincts about the kind of person that gets hooked. It helped, of course, that he reminded me of myself. He was white, male, neutral accent, like me. (And harm-reduction experts have long complained, and rightly, that journalists only recognized the depth of the opioid crisis when it began to kill people that resembled them. I'm caught in The Loop just like anyone.) But beyond that, he also talked about heroin in a way that I've come to understand is the most accurate picture of its power: as an irresistible force, like a storm front or a tide. In years of speaking with people in its grip, their feelings and their language on the subject are the same. It is cause and effect. When it rains, you get wet.

Like many people—maybe most people—I spent most of my life thinking of heroin as a mistake other people make: a poisonous vice whose marketing gimmicks somehow just work more strongly on a weaker kind of person. I figured the only folks who would fall into using it would be those without the right upbringing, those with an addictive personality, those who just can't control themselves. But as I listened to Patrick, who had a girlfriend and a place to stay in one of the most expensive cities in the world while carrying his terrible addiction, I realized I may not understand other people, or myself, as well as I thought I did.

My inflated opinion of myself, my utter blindness to my own chances of becoming addicted to heroin, the way I reassure myself about my place in life by ascribing Patrick's circumstances to some sort of failure of character—all of that is a terrible misjudgment, and also an outgrowth of a survival strategy humans have pursued for generations. Evolution has built us into beings with tremendously high self-regard and an enormous capacity to either ignore or rationalize in ourselves what we decry in other people.

First of all, we're incredibly upbeat. In 1969, a pair of researchers published the "Pollyanna Hypothesis"—the idea that human beings are simply more likely to remember positive information than negative information. A 2014 analysis of 100,000 words across ten languages confirmed that vastly more of the words we use describe good things than bad things.

And we like our own opinions. In the 1990s, the social psychologist Jonathan Haidt established the theory of "social intuition," which posits that people develop their moral intuitions about the world almost automatically, and then essentially reverse engineer a system of reasoning that supports those moral intuitions. His theory is that any conscious weighing of morality we do usually winds up being used to justify the intuitions we already had.

We're also self-congratulatory. In 2004, a meta-analysis of more than 266 separate studies firmly established what researchers call the "self-serving bias."[1] When something goes my way, my brain tells me it's because of who I am. When something doesn't go my way, my brain tells me it's because of some external factor beyond my control. (In people with depression and anxiety disorders, very often this sort of attribution is reversed—a terrible curse.) This is why my first instinct is to assume I haven't become addicted to opioids because I'm somehow of stronger moral character than other people, even though it's a scourge afflicting millions of Americans across the nation. But if I *were* to become addicted to it, I'd undoubtedly blame the drug, my circumstances, and bad luck. Meanwhile, you'd be more likely to consider my addiction a personal failure. And so on.

All of this makes perfect evolutionary sense. My wife and I often joke that perhaps there was once a branch of the human evolutionary tree that was great at remembering and talking about how painful it is to give birth to a child. "Wow, that was terrible!" we imagine one woman of that branch telling her mate. "Let's never do that again!" She and her kind are long dead, the joke goes. Only we upbeat amnesiacs are left thousands of years later. (A Swedish study of 1,300 women found that in fact a small number of women who experienced the worst kind of pain during childbirth remembered it quite well, actually. But in the rest of them, who didn't report the worst kind of suffering, the memory of the pain consistently faded over time.[2])

And so we wander around, optimistic, forgetful, self-serving. It's a wonderfully useful set of traits to possess, or at least it once was. But it gets us in trouble now. It means we have a sort of built-in immunity to accurately recognizing and analyzing our modern habits and cravings—our inability to put down the phone and get a good night's sleep, our opioid epidemic, our shared difficulties and vulnerabilities—and how they define our modern selves.

In the 1970s, psychologists began experimenting with the implications of our highly evolved feeling of self-confidence. The psychologist Richard Nesbitt identified, for instance, an "actor-observer asymmetry": a difference between how we explain our own behavior to ourselves versus how we explain the behavior of other people. I wish I'd known about it before meeting Patrick. According to Nesbitt's findings, when I make a mistake (*I forgot my house keys!*), I tend to blame that behavior on the situation I'm in (*It's been a long week, I've got my kids waiting in the car, anyone might do this*). But if I observe that behavior in someone else (*My wife forgot the house keys!*), I tend to blame it on that person's disposition (*My wife is so forgetful, my God, this is always happening*).

In the years following Nesbitt's paper, psychologists performed dozens of independent studies that supported his findings, until another psychologist named Bertram Malle surveyed all the studies and found that it's more complicated than that. The way we describe intentional behavior to ourselves, for instance—like the decision to buy a particular car—has to do with a complex web of reasons and beliefs and desires. But Malle did find that when it came to unintentional behavior—locking ourselves out of the car—the actor-observer asymmetry held up.

What behavior could more perfectly fit the actor-observer asymmetry than our national attitude toward drug addiction? There are enough opioid painkiller prescriptions written in the United States each year to give every adult their own bottle of pills. These pills have the same chemical effect as heroin you buy on the street. And that, along with a complicated combination of welfare reform, malpractice law, and predatory marketing, is why so many thousands of people find themselves unintentionally developing dependence. Then, just as unintentionally—perhaps they run out of the prescription, perhaps they can't afford the legal stuff any longer—they wind up in the grip of heroin or whatever pills they can score on the street.

And yet from parents to presidents, the dominant attitude toward drug addiction has been what I believed for most of my life: that it's a matter of character. But if we look at the science of habit formation and persuasion, that's clearly not the case.

OUR INTERNAL MECHANISMS of compulsion, along with the gauzy layers we drape over them to make them more attractive to our self-image, have created an extraordinary opportunity for taking advantage of our fellow humans by marketing to our involuntary decision-making systems and the ways we rationalize our involuntary choices. Those mechanisms are a powerful yet largely unexamined part of what makes capitalism go. And the top people in marketing and persuasion often start out as researchers studying those mechanisms.

In 1965, Robert Cialdini was an undergraduate at the University of Wisconsin, doing terrible things to earthworms. Then Cialdini's humanity kicked in. Not his empathy for the invertebrates themselves—he talks about his time torturing worms for science with the same enthusiastic intensity with which he describes the rest of his work—but another emotion. "I had a mad crush on Marilyn Rapinski at the time, who was taking a social psychology class, and there was an empty seat next to her, so I sat in on the class." Soon Cialdini was transfixed by the subject and went to graduate school to study it. In the end, his college crush lured Cialdini down a path that led to being published in thirty-six languages and traveling the world as an expert on how we persuade one another to do certain things.

As a graduate student Cialdini quickly grew tired of the same old methodology. "The way I'd been doing things with college sophomores on college campuses was not representative of the factors that make the biggest difference outside of that controlled environment. So I decided to go see what the professionals do." Cialdini went on a bizarre three-year journey into the maw of American chicanery. "I presented myself as an applicant to various sales training programs. I learned to sell used cars, portrait photography over the phone, encyclopedias door to door, that sort of thing." He began to discover that a lot of the tactics taught in these various programs were junk. But the fundamental principles that worked were consistent no matter the industry, and so he began to compile those principles.

His manifesto, *Influence*, remains one of the most-read business books of all time. It argues that six principles must be involved in any successful pitch: reciprocity ("Here, have a free sample"), commitment and consistency ("You've put so much into this process, don't back out now"), social proof ("This one is our best seller, everyone loves it"), authority ("JD Power says it's a best buy"), liking ("You have a lovely family"), and scarcity ("This two-door is the last one on the lot!").

Cialdini's work is devoured by readers in marketing and sales, and he is in constant demand as a consultant to multinationals, governments, you name it. Yet his work is also cited by academics studying human behavior and decision-making. A 2008 paper published in

Science explored the "broken windows theory" that assumes people observing signs of disorder are more likely to break rules themselves. The paper's authors, Dutch social psychologists, set up their findings by describing the well-established idea that messages asking people to refrain from an activity (like littering) are more effective in a setting where that thing hasn't already been done (a clean, unlittered park). "In honor of the individual who first described this process," the authors wrote, "we call this the Cialdini effect."

Even the patterns of Cialdini's *success* offer him useful information for analyzing his own work. Publishers in Thailand recently began printing his books. He has published papers in Polish psychology journals. His consultancy, Influence at Work, has worked on tax-collection strategies for the British government. His findings seem to be culturally universal. Well-established and young companies alike draw on his work. Articles about making social media more effective, extending the reach of email marketing campaigns, and improving sales numbers for online stores all regularly reference *Influence* and his six principles. "But when that book was written, there was no Internet, there was no e-commerce," Cialdini says. "That's what's validating."

I asked Cialdini whether he worries that he has inadvertently released into the wild what is essentially a user's manual for manipulating humans. He says he doesn't see it that way. According to Cialdini, he's attempting to inoculate us against these tactics. He says it's valuable to explain the work as widely as possible so that people are forewarned against manipulation.

But that doesn't seem to have been the effect, for two reasons. First, his buying audience isn't everyday people looking to protect themselves. It's people in marketing, an industry that doesn't need to read the peer-reviewed journals or think through the ethical implications to use our ancient wiring to sell modern products.

Second, and more important, being forewarned doesn't seem to do us much good anyway. I no longer drink, after a prolific twenty-five-year relationship with alcohol, and in spite of being acutely conscious of the fact that drinking for me sets off a cycle of poor sleep and low moods that ultimately pulls me into terrible depression, I still feel the attraction to booze whenever I'm around it. It's the sensory reminders of drinking that really do it to me. A dark bar, seated at the end where it wraps around and meets the wall, so I can watch the bartender work. The first antiseptic sting of a Manhattan on the rocks. Fishing the dark, boozy cherry out with a toothpick when the drink is finished. And this, I've learned, is why someone like me should essentially never set foot in a bar again.

The USC psychologist Wendy Wood once broke it down for me in front of a taqueria near her campus. Wood studies how we form good and bad habits, and she was trying to explain to me why I constantly return to the same order—carnitas burrito, guacamole, no sour cream—whenever I'm hit by the sights and smells of a taqueria. "We outsource a lot of our decision-making to the environments we live in," she told me. Once upon a time this wasn't a vulnerability, it was an evolutionary advantage. We could quickly escape the bear, spot the outsider, grab the fruit, by simply following the environmental cues around us. ("Growling! Stranger! Ooh, shiny berries!") It saved us enormous amounts of time and cognitive power and saved our lives as a result. But in modern life, that same circuitry sends ex-drinkers like me into a personal odyssey of self-evaluation every time I walk past the dark doorway of a bar. It's carrying all sorts of signals into your brain, too. And when the signals arrive—as emotions, instincts, things your "gut" is telling you—we evaluate *them*, not the merits of the actual decision.

In fact, science has begun to establish that our unconscious decision-making systems become that much more irresistible when we let our emotions make our choices for us. Back in the 1980s, Paul Slovic was approached by the city of Las Vegas and asked to find out whether potential visitors to Sin City would still be inclined to arrive in town ready to party and gamble if they knew, say, that the federal government had put the nation's largest repository of nuclear waste at Yucca Mountain, roughly eighty miles away.

"At first I thought we could ask people 'hey, if we build this thing, will you still come?' But I discovered pretty quickly that you couldn't trust their answers," says Slovic. Their reactions were instinctive, and emotionally driven, and not at all reflective of any rational analysis of risk. Essentially, he found that no matter what other reassurances were offered, if tourists were told that Yucca Mountain was nearby, the information poisoned the relationship. "If you mention nuclear waste and Las Vegas in the same sentence, they immediately say 'no, of course, I'll never visit again.'"

At the same time, Slovic had come into his trove of early documents from the tobacco industry, in which it described hiring scientists who determined more or less the same thing. "They were told by their consultants who were doing research, 'take all the rational reasons for smoking your brand out of there,'" he says. Instead, "make it all about good feelings, you know—in beautiful environments, doing exciting things with your cigarette." A positive emotional association with a particular brand of cigarettes essentially overrode any other, more analytical form of judgment. And a negative emotional association with nuclear waste overrode any reassuring information tourists might have had about Vegas.

Slovic believes that the ancient, instinctive system I use to make the same burrito order every time is in fact wonderfully useful. It got us here. It gets us through most of our day. "Most of the time, we're doing it all with our feeling and experiential system," Slovic says. "And it works pretty well." But it lacks something important, he says. "This very sophisticated system can't count." When presented with information about modern, abstract risk—the distance to Yucca Mountain, the statistical probability of developing cancer from smoking, things out of our line of sight—we're trying to use an emotion- and instinct-based system built for avoiding spiders and picking berries. And what's more, Slovic says, our ancient systems are easy to manipulate as more and more behavioral research like Cialdini's is adopted by business. "If they all become sophisticated in the System 1 and System 2 dynamics, which they clearly are," he says, "people will be using this for their own purposes, and our empathy, our feeling system will be hijacked."

WHAT ARE THE specific human instincts being used to manipulate us? What's the way in? As early as the 1970s, Kahneman and Tversky were working on a new concept for measuring human instincts around risk and uncertainty.

In their 1979 paper "Prospect Theory: An Analysis of Decision Under Risk," Kahneman and Tversky identified a tendency that upended decades of thinking about risk analysis. Rather than making rational decisions based on a thorough consideration of the straight probability of a good or bad thing taking place, they theorized, humans instead weigh gains and losses on a strange curve: our hatred of loss makes us wary of even statistically smart gambles, and we'll do almost anything to convert a tiny bit of uncertainty into a sure thing. This was the beginning of what Slovic would go on to show was an outsourcing of decisions to our emotions.

Kahneman and Tversky deployed a devilish sort of question to tease out this tendency.

> Choose one of the following gambles:
>
> (A) A 50% chance of winning $1,000
>
> (B) $450 for sure

They found that an overwhelming majority of respondents from across the world—Israel, Sweden, and the United States—chose the sure thing, although statistically the value of the first gamble works out to a higher payoff. And over the course of their careers, Kahneman and Tversky discovered that this tendency manifests itself in all sorts of other human irrationality.

Now, personally, I can relate to this hatred of uncertainty, and to being terrible at dealing with it. I'm a very terrible but very devoted surfer, and I'm often having to talk myself into a dark, cold patch of ocean that offers no visibility beneath the waves. My fears while in the ocean are various—drowning, being injured by the board, being contorted to the point of breaking my back in a wave, as has happened to a friend—but sharks are for some reason my most vivid fear. The incidence of shark attack in the United States is astonishingly small, with sixteen or fewer attacks per year, and at *most* two fatalities every two years. Even if I spend an hour or two in the ocean every day, my chances of being attacked by a shark are statistically smaller than were my chances of being crushed by a vending machine or choking to death on a champagne cork when I lived in New York City. And yet the few times a year I go surfing, my eyes will instinctively glom onto any sign of the circular upwelling of water that I've read indicates a column of water is being pushed ahead of a rapidly surfacing shark. I panic multiple times per session that something is about to grab me with a torso-sized maw full of knives and drag me under. And that's why for some reason I've developed an unconscious determination to never look under the surface. Whether I'm forced under by a wave or falling off the board, I close my eyes fiercely while I'm submerged and scramble for the board when I come back up, fighting panic. This is my brain's effort to negotiate with my fears, and to wipe out the uncertainty of my situation.

That horror of uncertainty, and the tendency to find reassurance in empty gestures, is now the basis of whole industries, as the economist Richard Thaler explained to me. Thaler won the Nobel Prize for carrying Kahneman and Tversky's prospect theory forward into new domains, formulating along the way notions like the endowment effect, the theory that when evaluating two identical items, the one that belongs to me is magically endowed with greater value in my estimation. Thaler originated, with the legal scholar Cass Sunstein, the notion of "nudges," small design features within a system that can gently guide us toward choices that benefit us. For instance, whereas a decade ago it was common practice to ask new employees to opt in to their 401(k), Thaler's work helped inspire a new standard: it's now common practice at most major companies to enroll employees automatically, which has hugely increased the numbers of people saving for retirement. (The work has been so successful that an entire governmental agency, a nudge unit, was created in the United Kingdom to study and implement tactics for improving British life.)

But Thaler despairs of the human allergy to uncertainty. We sat together in the atrium of the Booth School at the University of Chicago, where he teaches, and he began by explaining to me what he and others have found: that humans are hopeless with probability. "In fact, there's really only five probabilities the average human can handle," he said. "Want to guess what they are?"

I shrugged.

"They're 99 percent, one percent, 100 percent, zero, and 50-50. That's it." He laughed.

"And we'll do anything to push 99 percent to 100," he said, leaning toward an open laptop. We sat together scrolling through page after page of sites offering insurance on the various things one can buy through the Internet. We looked at a pendant, for instance, that cost only $34.99. And yet for $35.00, an outside company would insure it against loss or theft. "You could just buy another one!" Thaler said. "That's got to be the dumbest of dumb insurance."

But what particularly irks him is travel insurance. In a 2015 *New York Times* op-ed, he railed against the fact that airlines like United routinely require us to actively turn down travel insurance *even when buying a fully refundable airline ticket*. And in a larger sense, he railed against the idea that these dark nudges—what Robert Shiller and George Akerlof called phishing in their 2015 book *Phishing for Phools*—are being deployed by the private sector. Critics of his and Sunstein's work have complained that government policies that nudge us to make better decisions are paternalistic, maybe even unconsciously dictatorial. But, he wrote, one has a voice in what a representative government does. But if a manipulative tactic makes a company money, not only can't we vote that company out of office, its competitors are likely to copy it.

This blindness to probability, combined with our inability to grasp the true risks of a situation, means we're unaware of the real risks we're taking when we put our heads underwater.

I ENCOUNTERED THAT dynamic most vividly in 2021 when I spoke with Kathleen Wilkinson. Wilkinson lived for forty-three years in Prairie du Chien, Wisconsin. It's a small town, popular with tourists who come for the hunting and fishing, or perhaps for the Father's Day weekend Rendezvous, full of buckskin costumes and antler mugs, that reenacts Prairie du Chien's history as a fur-trading camp. But that town, where the Mississippi meets the Wisconsin River, was not a happy place for Wilkinson. She survived a toxic twenty-four-year marriage there. And she suffered a spinal injury in the course of her job as a medical technician in 2016, which isolated her further. Finally, as she watched her mother's home go to pay the gambling debts of a family member, she decided she had to get out.

Wilkinson's exit was accidentally paved by her first husband. He created a profile for her on a dating website, in a perverse act of digital control, and one day she got a message that read simply "hello."

"There was a picture of this cute blond-haired man in a wheelchair, with blue eyes, surrounded by snow," Wilkinson remembers. She wrote back, and soon they were in regular contact. "That man is now my husband."

Today Wilkinson lives in the mountains of Montana, in Kalispell, on the edge of Glacier National Park, on property she shares with her new husband and his father. It's been a paradise, she says, but a dark compulsion followed her from Wisconsin. Back in Prairie du Chien, while she was laid up with her spinal injury, she got a pop-up on Facebook for a game. "It looked like fun, so I tried it," she recalls.

The game was what people in the business call a social casino app, one that combines the classic casino experience—slots, poker, blackjack—with social activities like text, audio, and video chat between players, and often the ability to form teams and clubs. The companies that make these games are often the same companies—or are owned by the same companies—that own and operate real-world casinos. Early studies of the social casino business found evidence that people who play the games were sometimes transitioning to real-world gaming,[3] making it a logical "customer acquisition" tactic on the part of a traditional casino business. But in the end, the companies didn't need to bring these players to a real casino. It turns out social casino games can become deeply habitual and costly to people who would never physically walk up to a slot machine or a card dealer. And as the pandemic set in, these companies discovered that while their real-world casino machines were suffering, the apps were thriving.

Wilkinson was already in over her head by the time the pandemic hit. She remembers telling her husband that she thought she might have a gambling problem. She simply couldn't stop playing this little game. But you're not playing for real money, he reassured her. That's not gambling.

Experts in addiction and machine gambling could have told her the truth, if only she'd had one around to ask, not that it likely would have made a difference. Natasha Dow Schüll, an associate professor at NYU, spent twenty years with players of slot machines, with executives and designers at the companies that make them, and with regulators struggling with the right legal framework for analyzing them, to write her 2012 book *Addiction by Design: Machine Gambling in Las Vegas*. "What I learned over my many years of fieldwork," she says, "and it took me a long time for it to really sink in, is that the gamblers are not there to win money, they're there to spend time." If anything, winning an unexpected amount of money can detract from their experience. "Gamblers used to tell me, 'I get irritated and frustrated and sometimes angry when I win a jackpot.'"

She came to realize "they were not there to win a jackpot necessarily, but to keep going. So what they're sitting there for is the kind of escape that they describe as 'the machine zone.' It's a zone where time falls away, money falls away. And you just keep going."

That zone, she's found, is the basis of the gaming industry. Once upon a time, slot machines were built as a quick distraction, a place where you'd throw in a quarter and walk away, either a little richer or feeling a little foolish. "Old slot machines didn't even used to have seats in front of them," Schüll points out. "Not so today's games. They've got ergonomic seats, and they're really meant for you to sit and spend a ton of time at and not realize that you're losing money as you're sitting there."

There is something in human programming that predisposes us, to varying degrees, to fall into the zone, she says. "I have talked to very rational people who show me their bank statements. And there will be a whole page of withdrawals, twenty minutes, two hours, the next thirty minutes of $100, $20, $20, $40. And clearly, that otherwise rational person is thinking something new each time they take money out, and not learning that this isn't the way to go." And, she says, the industry has a gauge for that. "Time on device," or T.O.D., has become the way machine-gaming companies measure success. "T.O.D. really is the revenue metric of the industry."

Schüll points out that if we want to begin quantifying the effect these products are having on the human mind, the data is already there—we just need subpoena power to get it. And there are signs that it could happen. In Massachusetts, for instance, a 2011 law that permits commercial casino operators to open three casinos and one slot parlor in the state has an often overlooked provision buried inside it. Section 97, which Schüll submitted, and which lawmakers, to her immense surprise, adopted, reads, in part,

> Gaming operations shall supply the Massachusetts gaming control board . . . with customer tracking data collected or generated by loyalty programs, player tracking software, player card systems, online gambling transactions and any other such information system. . . . The commission shall convey the anonymized data to a research facility which shall make the data available to qualified researchers.[4]

One point of optimism for fighting back against behavior-shaping technology, that we'll see again and again moving forward, is that if companies like these are using data to prey on us, then much of what we need to know about our tendencies and how they show up inside a company's algorithms and tactics is there for the taking, in court. If lawmakers actually forced gaming companies to provide the data they use to track player behavior, Schüll says, we'd know just how well they understand the effect they're having on the minds and decisions of the players at the slot machines and card tables. "It's absolutely the case that you can see the signature of addiction in the data that is being so robustly collected by these companies."

How does she know? Back in 2011, Schüll discovered a Canadian company, iView, which took an algorithm built from two and a half years' worth of player data from casinos in Saskatchewan and used it to detect problem gamblers in real time. The system uses all the means by which a casino would otherwise ensnare someone in the grip of compulsion to try to rip them free from it. The iView software records five hundred variables in player behavior, from the number of machines played to the days of the week, and calculates a risk score. When that score is combined with on-the-floor observations by trained casino staff, the system has been found to be 95 percent accurate in spotting people who are no longer in control of themselves. At that point, all advertising to that player freezes, the player's funding card freezes, and the casino's facial-recognition system tags them to casino managers as a risk alert.

It is easy, sitting as you probably are, somewhere far from a casino, to shake your head at all of this and lament the poor souls who have lost so much on the gaming floor. But what Schüll and others have been trying to communicate for so many years is that there is no difference between an addiction to slots or poker and an addiction to a game you play on your phone.

The companies that make those games would, and do, argue that there's an immense difference: gambling for real money is a highly regulated activity. Playing a simulator that offers no chance of winning real money is not considered gambling.

And the game that Kathleen Wilkinson was playing, DoubleDown Casino, didn't offer the chance to cash out. With social casino games, money only flows one way: *into* the game, as it offers the chance to pay for expanded play time, for the chance to communicate with friends, for higher tiers of play. When she began playing it, Wilkinson considered it a diversion during a lonely phase of her life, and the $2.99 she spent here and there to connect with friends as coplayers and access new games seemed like a small price to pay for a little fun and virtual company.

By the time she told her husband that she was concerned about her compulsive behavior, she'd lost track of how much she was spending on the game. But when they looked more closely, she realized she'd spent roughly $50,000 on the game over four years. Wilkinson had planned to retire into a life of caring for her husband in Montana. But she now realized she'd have to work, with a spinal injury, well into the last phase of their life together. "I wish to God I had not spent that money. Because I'm worrying about being able to take care of my husband and I have no income coming in right now," she told me, fighting tears. "I just wish they would do the right thing and give back the money."

Wilkinson is now part of a class-action lawsuit against DoubleDown Interactive. When I describe this effort to most people, they laugh, and mock the plaintiffs for being gullible. "People who wasted fortunes on those apps have the problem, not the app companies," a retired physician wrote to tell me after I published an interview with Kathleen Wilkinson.

But it turns out there are a *lot* of people like Wilkinson, and the companies have been doing an uncanny job finding them. In fact, according to Wilkinson's lawyers, there are thousands of people in the United States who have spent at least $10,000 and as much as $400,000 on these games. "These social casinos are targeting people who are most likely to be addicts, and then taking their life savings," says Jay Edelson, Wilkinson's attorney. "These aren't rich people. This isn't Michael Jordan going to a casino losing $2 million and it's a bad night and he moves on. These are ordinary, hardworking people who've saved up all of their money, they're ready for retirement. And then they get sucked up into this world that nobody really talks about."

I've seen that world, and the kind of thinking that built it, first-hand. Here's what that looks like.

A FEW MONTHS after Patrick described his heroin overdose to me, I was invited to join a loose dinner party elsewhere in San Francisco thrown by a group that called itself BTech, short for Behavioral Tech. It was a monthly gathering of young neuroscientists, behavioral economists, and other specialists in human decision-making who now work in the tech industry. The gathering of perhaps fifteen or so that night included someone from Lyft; someone from Fitbit; the creator of a financial-advice chatbot; and Nir Eyal, a Stanford MBA who had sold two companies, written a best-selling book called *Hooked: How to Build Habit-Forming Products*, and was now a consultant successful enough to have built out the white-walled former storefront in which we were all seated. I wrote to the group ahead of time to inform them that I'd be there, that I was writing a book, and that whatever they said might wind up in it. Over Coronas and Indian food, we listened to a pair of newly graduated neuroscience PhDs, T. Dalton Combs and Ramsay Brown, talk about their company, called Dopamine Labs, through which they were selling their academic knowledge of dopamine release to help make apps more addictive. They were there that night to talk about the market potential, and to offer their services.

Combs and Brown began the evening showing off an app they'd built, called *Space*, which could interrupt the feedback loop one gets from opening social media apps by enforcing a few seconds of mindfulness before the app would open. But by the end of the evening, it was clear they were happy to sell these same principles in the service of almost any company. As Combs told us, what's so great about the human mind is that if you can just manipulate someone's habits, their consciousness will invent a whole story that explains the change of behavior you tricked them into, as if it had been their idea all along. It makes what they were calling "behavior design" so easy, he told us.

Now, the room, I should point out, was full of good intentions. Many of the apps had to do with fitness and saving money and healthy meal plans. But there was an open way of discussing neuroscience and behavior change and habit formation that felt like a loaded gun. And when one member of the group asked the Dopamine Labs founders whether there was any company they wouldn't be willing to work for, Combs said, "We don't think we should be the thought police of the Internet."

Dopamine Labs' presentation centered on the idea that companies can use our hardwired habit system—the same stuff studied by the likes of Kahneman, Tversky, and Slovic, what the two neuroscientists described as a chain of triggers and actions and feedback and rewards that can be manipulated the way one might pull levers to drive a crane—to get people to turn over information about themselves through apps. These were folks trying to understand how to manipulate—or, hell, *create*—the habits of their customers. And eventually we wound up talking about drug addiction.

At one point in his presentation, Combs said that humans form habits through contextual triggers. "If you take a habituated drug user who has recovered, let's say it's someone who used to do cocaine in nightclubs—" He checked that we were with him. Forks and beer bottles were frozen mid-journey. "—Let's say you bring them to a nightclub, and you show them baking soda, and you even tell them it's baking soda, they'll still want to snort a line of baking soda." (That's me outside the taqueria, I know now. That's me inside any dark, wood-paneled bar.)

Someone in the room asked whether everyone is equally vulnerable to drug addiction. I felt relieved that it had come up. After all, we'd been talking for nearly three hours about the universal mechanisms of the human mind and how to take advantage of them. At this point there seemed to be a collective realization that we were playing with live ammunition. We'd clearly begun to push against the ethical boundaries of all this behavior design. But somehow the people in the room at that point managed to elevate themselves above the everyday people they imagined would use their apps, drawing distinctions that conveniently left those of us gathered in the room above the water line. And Eyal jumped in with a sweeping statement. "Let's be honest: the people in this room aren't going to become addicted. If you were injected with heroin a hundred times"—he gestured at Combs, and then at the rest of the room—"you're not going to become addicted to heroin." He seemed utterly convinced of this. "Unless you've got some sort of deep-seated pain in you," he says, "it's just not going to happen."

I didn't share my opinion that pretty much any mammal injected with heroin one hundred times would likely become addicted, but I did mention to him that opioid overdose kills more people than car accidents these days, that at that time it was the number-one cause of death by accidental injury in the United States, that people from all walks of life were falling into opioid addiction at a rate never seen before in American history. But he was unshakeable. "The people who become addicted have pain in their lives. They return from war, or have some other trauma, and that's what gets them hooked," he told me calmly. "Only two to five percent of people are going to become addicts." Forks and bottles began moving again.

Combs and Brown went on to appear on *60 Minutes* a few months after I met them, and shortly after that they renamed their company Boundless Mind. Soon Arianna Huffington's Thrive Global bought them, and Combs was their head of behavioral science until October 2020. He now runs a metabolic-fitness company. Dopamine Labs does not appear on his LinkedIn profile.

Eyal went on to write a book called *Indistractable: How to Control Your Attention and Choose Your Life*. It seems like a dramatic reversal for a guy who wrote a marketing handbook called *Hooked*, but it's not. His thesis in *Indistractable* is that it's up to individuals to learn self-discipline, that technology is not addictive, that self-control, not regulation, is our pathway forward. This is someone who used his understanding of psychology to establish and sell large-scale tactics of persuasion and habit formation. And yet he had somehow convinced himself—and, seemingly, most of the people there that night—that the people making this tech are somehow not only able to resist what we had just spent the evening establishing is an eons-old tendency of the brain, but that it was all right for that rarefied group to deploy products that played on those unconscious tendencies in the rest of us. Eyal went on to tell Ezra Klein in a 2019 podcast appearance, "The world is bifurcating into two types of people: people who allow their attention to be manipulated and controlled by others, and those who stand up and say, 'No, I am indistractable. I live with intent.'"

When I watched the presentation from Dopamine Labs and heard Eyal's theories that night, I hadn't yet met Robert Cialdini or Paul Slovic. I hadn't yet met Wendy Wood, and she was still a year away from publishing a 2018 review of habit-formation science with the business professor Lucas Carden, in which they looked across the field and concluded that people are not in control of their choices. Wood and Carden compared our vulnerability to forming unconscious habitual associations (between the smell of a grill and ordering a burger; between the feeling of taking our work shoes off at the door and the first pull of an after-work beer) to the inexorable pull of a rip current. The trick, they wrote, is to not get in the water. "We now know that self-control involves a wide range of responses beyond willpower. To be successful, people high in self-control appear to play offense, not defense, by anticipating and avoiding self-control struggles."[5]

And when I sat there with those young entrepreneurs, earnestly talking through the ethics of addiction and design, I hadn't yet met

Kathleen Wilkinson, who lost herself in an innocent diversion, costing her and her husband their last years together. I hadn't seen the ways that our most ancient instincts were being built into emotion-shaping and decision-guidance systems for vast profit. I hadn't yet discovered that our unconscious tendencies are being amplified and shaped and mutated for profit, by accident and sometimes even on purpose. But even back then, I was frightened by the conversation I witnessed that evening, and, knowing what I know now, I still am.

Chapter 6

LIFE ON THE RAILS

WE'VE SEEN HOW the first loop works, the unconscious decision-making that drives so much human behavior, and we've begun to look at the second loop, the ways that technology-driven businesses are working to sample our unconscious decisions and toss them back at us. As part of that, I want to step back and look at how we not only came to pursue these sorts of manipulative technologies, but why we so readily believe in them and trust the guidance they offer.

The truth is that in our history we've come under the direct supervision of guidance systems before. And in the earliest iterations of them, they came to us in the guise of lifesaving technology.

Between the end of World War II and the 1980s, organizations like the RAND Corporation, the Cowles Commission for Research and Economics, and Stanford University's Center for Advanced Study in the Behavioral Sciences (where I spent a year as a fellow oblivious to this part of its history) attracted bright young thinkers from universities across the country who spent their time on the most pressing mission of the era: how to avoid nuclear war. In their brilliant book *How Reason Almost Lost Its Mind*, authors Paul Erickson, Judy L. Klein, Lorraine Daston, Rebecca Lemov, Thomas Sturm, and Michael D. Gordin describe a long, strange period during which these thinkers behind "Cold War rationalism" held military leadership, academia, and political theorists in their thrall. As we face the prospect of a world run by pattern-recognition technologies, we should closely study this period.

It began with a triumph of logistics over death itself. In the spring of 1948, American, French, and British forces, each occupying an agreed-upon zone of defeated Berlin, were locked in a spiral of growing tension with the Soviet forces there, who watched their Western counterparts with increasing alarm. Under the terms of the Yalta and Potsdam conferences, Berlin fell deep inside Soviet territory, yet the city was divided four ways. The Americans and British announced plans to merge their zones, with the French soon to follow. Then the Allies announced a new currency for their portion of the city, which would devalue the Soviet-backed currency, and that's when the Soviets imposed a blockade of air, rail, and road traffic to the Allied zones, cut off their electricity, and waited for starvation to drive the Allies out of Berlin.

But on June 26, 1948, RAF and USAF planes under the joint command of the Americans began to roar in and out of the Gatow and Templehof airports in West Berlin, packed with food and supplies. These were the first of nearly 300,000 flights that filled the sky above Berlin for more than a year. And because the pilots followed disciplined procedures, remaining at designated altitudes, navigating precisely mapped corridors back and forth, and taxiing with rigorous precision, they managed to form a lifesaving, uninterrupted circulatory system for the entire city. The flights and the unloading and loading were so efficient that at one point aircraft were landing in

Berlin every sixty-two seconds. The Soviets had assumed that without trains and trucks the Allied half of Berlin would starve. Instead, within eleven months the USSR had agreed to lift the blockade, and the foreign ministers of the four nations were sitting down together.

The Berlin Airlift, a revolutionary rules-based system that had saved an entire city and fought off the influence of the Soviet Union, set Cold War rationalism in motion. This new institutional enthusiasm sprang from a movement inside the US military, one accelerated by World War II, to apply computing resources and data analysis toward creating new efficiencies and strategies. Mathematicians at the Pentagon had already formed the Project for the Scientific Computation of Optimum Programs, or Project SCOOP, tasked with using the punch-card technology of the time to build new, data-driven systems. As Erickson, Klein, and their coauthors describe it, the Project SCOOP team managed "the ingenious design of an algorithm for deriving economically rational decisions from an equation system that included an objective function maximizing, for example, the tonnage delivered to Berlin." But because they didn't yet have the computing power necessary to execute such an algorithm, "the only optimal program that Project SCOOP could determine was for the least cost diet that could provide the essential nutritional needs of an active, urban economist weighing seventy kilograms."[1] The system had to work toward the simplest possible goal, which was the calorie needs of the average man inside Project SCOOP—a primitive but satisfactory objective function.

Inspired in part by Project SCOOP and its success in Berlin, a new field of rationalism exploded. Researchers like the economist Herbert Simon, the mathematician Anatol Rapoport, and the economist-turned-arms-control-expert Thomas Schelling began their careers studying organizational psychology. But geopolitical forces and military funding pulled the field toward questions of how entities like the

United States and the Soviet Union might avoid escalating tensions to the point of mutual annihilation. What those researchers were hoping to accomplish, it seems, was a sort of deprogramming—or maybe even preprogramming—of international relations, using formulas adapted from experiments like the Prisoner's Dilemma, in which two participants must decide, in isolation, whether to cooperate with or betray one another. The Cold War rationalists appeared to believe that with disciplined analysis and creative experimentation, it might be possible to boil down into a predictable set of patterns all the complications that gave rise to international incidents like the Cuban missile crisis. Understand those predictive patterns, and one could write the choreography necessary to avoid them. Erickson, Klein, and their coauthors describe the field as both admirably ambitious and dangerously naive.

But when I look at what those rationalists were after, I don't blame them for their naiveté, the way the book's authors seem to. Instead, I think they were limited by their time and their tools. What if those same researchers were recruited into a grand national effort today? They wouldn't be working for RAND, or even for DARPA. They'd be working at Facebook, or Google, or Amazon. And while the rudimentary flowcharts and columns they used to apply theoretical systems like the Prisoner's Dilemma look facile today, those were the same instincts—to find algorithms that could capture, predict, and even shape human behavior—that created AI, social media, and Big Tech. They just didn't have the horsepower or the fuel supply we do today.

"No rule, whether of logic or probability theory or rational choice theory, mechanically applied, is likely to be able to handle the gamut of political choices," Erickson and coauthors wrote with some frustration, and on this we agree. But time and again, we humans fall in love with logical-looking systems the same way Cold War rationalists did. We can't resist the temptation of offloading hard decision-making, and it's happening now more than ever. As we'll discuss further, researchers inside for-profit companies today are deploying the same reductionist instincts, along with huge amounts of computing power and algorithmic sophistication, to study what we choose to do in our lives. And while the stakes may not be as high as nuclear warfare, they are still frighteningly high. Because while the mind reels from the threat of annihilation by world-destroying weapons, it can't access its instantaneous snake-detecting, action-taking System 1 when faced with far-off threats. Even climate change, which threatens us as a species, isn't something we seem to *feel* until we're directly faced with flood or fire. And AI is a far subtler threat than that. In fact, for all the reasons we've discussed, we're primed to accept what AI tells us. And that acceptance is already getting us in trouble.

JUST AFTER 5 p.m. on April 9, 2017, passengers on United Airlines flight 3411 sat packed together at the gate at Chicago O'Hare, waiting for the plane to push back and begin its journey to Louisville. But the plane didn't move. Instead, a voice came over the cabin speakers to explain the delay. Four crew members needed to get to Kentucky to staff a flight there. Would any volunteers be willing to get off this plane in exchange for $400? The next flight wouldn't be for another twenty-one hours, though, and with their bags already smashed into place over their heads, the passengers were in no mood for compromise, and ignored the plea. The offer was raised to $800. No takers. And so the voice returned, this time announcing that four names would be selected, and those people would have to leave. A moment later, flight attendants approached a young couple who dutifully rose from their seats and trudged forward off the plane. But when the attendants approached Dr. David Dao, he didn't rise. Approached by a United supervisor, Daniele Hill, Dao identified himself as a physician on his way home, and he pointed out that missing tonight's flight would mean missing his rounds in the morning. He had patients depending on him to be there. He could not get off this plane.

Tensions rose. Hill was adamant that Dao had to leave. Dao was adamant that he could not. According to other passengers, Hill told Dao, "I'll have to call the police and have you escorted off the plane." And that's what happened.

As a hub of two major airlines, Chicago O'Hare moved nearly 80 million passengers that year, and the incredible volume of humanity that passes through the airport means not only that it's run by its own city agency, the Chicago Aviation Department, but that the department has its own police force. The roughly three hundred officers of the Chicago Aviation Police were never supposed to be full-fledged police: forbidden to carry firearms, they had only limited authority to make arrests, and in those cases their training was to hand suspects off to Chicago PD. But they were trained like police and had to meet roughly the same physical and educational requirements.

That night in April, Chicago Aviation Police officers James Long, Mauricio Rodriguez Jr., and Steven Smith walked down the aisle and told Dao to get up. Seated in the place he'd paid for, Dao refused. And then, tearing through the fabric of corporate-customer relations we all assume protects us, they put hands on him. Gripping him as he struggled, Long, according to another passenger, yanked Dao sideways, and the doctor's head bounced off the metal armrest. He went limp, and they dragged him by his arms past hundreds of gaping passengers. Phone video of the incident showed Dao visibly bloody and unresponsive. The United crew needing to reach Louisville got on the flight, and it departed, arriving two hours late.

The next day, as footage of the incident began to circulate on social media, United CEO Oscar Munoz put out what is today considered a famously disastrous piece of corporate communications, claiming in a statement that Dao's removal was necessary for "re-accommodating the passengers." An internal memo to United staff criticized Dao and praised the actions of Hill and her team. But within a week, as heat built on social media and United's share price began to fall, Munoz was apologizing to anyone who would listen. "No one should ever be mistreated this way," he told one interviewer.

Two weeks after the incident, on April 24, Dao announced through his lawyers that he would sue, claiming a broken nose, missing teeth, a concussion, and damage to his sinuses that required surgery. Three days after that, his lawyers reached an undisclosed settlement with the airline.

The story, in its endless retelling, has become a parable about bad choices under stress—bad choices that radiated upward from the United supervisor Hill to the CAP officers to United's CEO. But *why* did they make those choices? I believe they were caught in the loop. Not just the first loop, the species-wide one made of our most ancient instincts. The second loop, where those ancient instincts are played upon by new, invented systems.

After the passengers had steadfastly ignored several offers of money and hotel rooms, when the United staff could not cajole anyone to give up their seats, as the crew for the unstaffed flight waited with their bags at the gate, it was not at all clear how to break the stalemate. So a nonhuman system was invoked. Names will simply be chosen, the staff told the passengers. And suddenly everything changed.

Until that point, the situation was a human-to-human, transactional one, in which one thing of value (a voucher for a free ticket) was being offered in place of another thing of value (the chance to fly home and sleep in one's own bed). The payment from the passengers would be their inconvenience. As a transaction, this portion was transparent: why the deal needed to be made, what was on offer, what the cost would be. But the company wouldn't pay enough to make the voucher-for-overnight-stay mechanism work. With seemingly no other options, United's people on the ground ended the transparent human negotiation and activated an opaque system by which an unseen machine would simply choose which passengers had to get off the plane.

It turns out that when humans are subjected to a system they don't understand, they abandon many of the critical faculties they otherwise bring to bear in a more human interaction. There is an enormous amount of scholarship about this habit of mind, stemming mostly from anthropomorphism, the tendency to credit a confusing system with sophisticated attributes it does not possess.[2] Carey Morewedge, a professor of marketing at Boston University who studies decision-making and bias, told me, "Our findings suggest that when the domain has uncertainty, people tend to ascribe intelligence to explain outcomes in those domains." In this case, the passengers on the airplane, informed that names would be simply chosen, were suddenly at the mercy of a system they didn't understand, and their anthropomorphism was activated. If the airline chose those people to get off, passengers presumably reasoned, they must be the people who have to get off.

And other researchers have found that System 2, the one Kahneman and many others have shown is a delicate, inefficient, but extremely important error-correction system for our automatic System 1, is also easily distracted by nonsensical but convenient assumptions. The psychologist and University of Chicago professor Jane Risen studies decision-making through topics like superstition, going with one's gut, and other forms of magical thinking. In a 2016 study, she found that these tendencies are often so compelling that "people cannot seem to shake them, despite knowing that they are incorrect.

In these cases, people successfully detect an error of rationality, but choose not to correct it." Sports fans rationally know they have no role in the outcome of their team, no matter what they do in their living room at home, she writes. Half of Americans know there's no rational reason to knock on wood for good luck, but nonetheless admit they regularly do so. Our rational mind hands over the steering wheel: "System 2 acquiesces to a powerful intuition. Thus, in addition to being lazy and inattentive, System 2 is sometimes a bit of a pushover."[3] When an outside system is entirely opaque, and also plays into our magical thinking, as demonstrated by the passengers and crew and airport police on the United flight, everyone except David Dao, we do indeed get pushed over. Even in the face of patients needing a doctor or a man violently dragged down an airplane aisle, we find a way to trust the logical-looking system.

WE'VE BEEN PUTTING enormous, irrational trust in machines for nearly as long as we've been communicating with them. In 1964, not even a decade after McCarthy and his group gathered to theorize about thinking machines, a man built a simple conversational program that elicited powerful confessions from strangers. The German computer researcher Joseph Weizenbaum, working at the MIT Artificial Intelligence Laboratory, named his system ELIZA, for the Pygmalion character who learns to speak "properly." It could read text as it was composed by a human subject on a typewriter, analyze it, and then kick back sensible responses according to a rough script. Today, we'd call the thing a chatbot.

But while companies deploy chatbots today to stand in for simple, upbeat conversation in everything from customer service to job interviews, Weizenbaum was essentially just messing around. As he wrote later, "For my first experience I gave ELIZA a script design to permit it to play (I should really say parody) the role of a Rogerian psychotherapist engaged in an initial interview with a patient." Why Rogerian therapy? Because it's a style of interaction, popular at the time, of drawing patients out by reflecting their statements back at them. Statements like "I'm so depressed" are met with "Why do you believe you're depressed?" And so on. For Weizenbaum, it was the perfect conversational application for what he'd built. He began asking people to test out his system without explaining what it was or how it worked, only that one should sit and type to converse with it. The following was an exchange between a young woman, who speaks first, and ELIZA:

Men are all alike
IN WHAT WAY
They're always bugging us about something or other.
CAN YOU THINK OF A SPECIFIC EXAMPLE
Well, my boyfriend made me come here.
YOUR BOYFRIEND MADE YOU COME HERE
He says I'm depressed much of the time.
I AM SORRY TO HEAR YOU ARE DEPRESSED
It's true. I am unhappy.

The ELIZA program knew not only how to reflect back the last sentiment expressed, it could also hang on to certain notions and circle back to them—a simple piece of programming that happens to impersonate one of the best traits of a good listener.

My father is afraid of everybody.
WHAT ELSE COMES TO MIND WHEN YOU THINK OF YOUR FATHER
Bullies.
DOES THAT HAVE ANYTHING TO DO WITH THE FACT THAT YOUR BOYFRIEND MADE YOU COME HERE[4]

Weizenbaum wrote later that he built ELIZA to learn what he could about natural-language processing, but wound up discovering far more about human tendencies. And in the end, those tendencies caused him to flee the field. He spent the rest of his life critiquing the sort of work he'd been doing, and he died in 2008 after a long period of environmental activism. In his 1976 book, *Computer Power and Human Reason*, he described a series of shocks that had him rethinking his project and recoiling from its implications. Shortly after creating the first functional prototype, for instance, he recalled asking his secretary at MIT to test it out. She had observed him working on it for months, and presumably knew it was a computer program. And yet "after only a few interchanges with it," he wrote, "she asked me to leave the room." The conversation had become too intimate for her to let her boss eavesdrop. Weizenbaum knew that people form relationships with tools, cars, all manner of machines, but "what I had not realized is that extremely short exposures to a relatively simple computer program could induce powerful delusional thinking in quite normal people."

And that delusional thinking went far beyond Weizenbaum and his secretary. Weizenbaum discovered that experts in human behavior were only too interested in the possibility of robot-enhanced therapy. (In fact, I have several times described to business-school and engineering students Weizenbaum's ELIZA work and the ethical dilemma it posed for him. The reaction is largely shock that he walked away from so promising a business model.) The Stanford psychiatrist Kenneth Colby wrote about using something like ELIZA to supplement the job of human mental-health workers. "If the method proves beneficial," Colby wrote in a 1966 paper, "then it would provide a therapeutic tool which can be made widely available to mental hospitals and psychiatric centers suffering a shortage of therapists."[5] In the 1970s, Colby expanded on the ELIZA concept by building PARRY, a software simulation of a paranoid patient (used to train student therapists) that was indistinguishable from a human patient to most psychiatrists—it was the first piece of software to pass the Turing Test, which evaluates whether a person can tell the difference between a robot and a human in structured conversation. By the 1980s, Colby had sold a natural-language psychotherapy program called *Overcoming Depression* to the Department of Veterans Affairs, which distributed it to patients who then used it without direct supervision from a therapist. Weizenbaum was alarmed to discover that even Carl Sagan, the most visible spokesperson for the wonders of science, was really into the idea. "I can imagine the development of a network of computer psychotherapeutic terminals," Sagan wrote about ELIZA in 1975, "something like arrays of large telephone booths, in which, for a few dollars a session, we would be able to talk with an attentive, tested, and largely non-directive psychotherapist."[6]

Computer Power and Human Reason was the beginning of Weizenbaum's new public life as a critic of artificial intelligence, and, more broadly, of our willingness to do what it tells us. In a 1985 interview with MIT's *The Tech* newspaper, Weizenbaum described the whole concept of computing as "a solution looking for a problem."

> People come to MIT and to other places, people from all sorts of establishments—the medical establishment, the legal establishment, the education establishment, and in effect they say, "You have there a very wonderful instrument which solves a lot of problems. Surely there must be problems in my establishment—in this case, the educational establishment, for which your wonderful instrument is a solution. Please tell me for what problems your wonderful instrument is a solution."[7]

Flying from one place to another on a passenger plane is a string of encounters with large machines we do not understand. From the ticketing process to security checkpoints to the strange mechanisms of the plane hidden beneath the smooth plastic of the cabin, we entrust our bodies to a long series of mysterious systems. And everything in that system is designed to heighten our trust: the scripted language of the flight attendants and the captain over the intercom, the uniforms, the regimented timetables. It's extremely rare for the choreography to be interrupted the way Dao's journey was.

Think of what Weizenbaum might have said had he been seated on David Dao's flight. As the voice asks for volunteers, he might have leaned over to the person next to him and pointed out that the passengers are primed to believe that things will go as they always have. They're not ready for a departure from the typical experience. And they're not ready to think critically about the system and how it works. And so when a passenger is told that the airline is choosing people to be removed from the plane, Weizenbaum might have warned, as he did his fellow scientists in a 1976 essay, the passenger is all too ready to go along with the program. "He is rather like a theatergoer who, in order to participate in and understand what is happening on the stage, must for a time pretend to himself that he is witnessing real events," he wrote.

That certainly seems to have extended to the aviation security police officers summoned to pull Dao off the flight. James Long, who went on to sue the Chicago Department of Aviation for firing him from his job as an airport police officer, said in his complaint that United flight attendants "knew or should have known that calling the Aviation Police . . . to remove a passenger who was refusing to leave their plane would require the use of physical force." He also blamed the Chicago Department of Aviation's "negligence and failure to train Plaintiff how to respond to an escalating situation with an Airline Passenger."[8]

Long was widely criticized in news reports for what looked from the outside like a desperate attempt to deflect blame through litigation. But I'd argue that Long and his fellow officers were in a sense helpless to resist the same sort of anthropomorphic pressures that Weizenbaum helped to discover through ELIZA back in 1964. They were all caught in the loop, just like the passengers and the gate agents and the crew. No one in the chain of decisions on and around that 2017 United Flight—no one except Dao—was using their full human faculties. (Although in my view, it's the United workers that most fully abandoned their senses—they summoned police and sicced them on a passenger.)

But one thing we learn from this episode is that when you hand difficult decisions—who is going to go home and who is going to be stranded in Chicago—to a mechanism no one understands, everyone gets dumber. Everyone—from the flight attendants who insisted Dao change flights despite the consequences for his patients, to the young couple that got off when asked, to the officers summoned to remove Dao—was acting under the direction of a larger, mysterious machine to which they'd all to varying degrees ascribed sophistication it did not possess. And although United Airlines has never explained exactly *how* Dao's name was chosen, its contract of carriage says that several factors (including the class of ticket purchased, whether there's a connection to be made at the next airport, frequent-flier status, and how recently the passenger boarded) go into algorithmically choosing who gets kicked off first.[9] A black box, if you will.

We've all obeyed a system like this: a FICO score, the fire claxon in our building when we know it's a false alarm, the dashboard light telling us to bring the car to a mechanic. We just don't consciously see that it's becoming more and more common to do so, as the modern world builds a secondary loop of control inside which we're all traveling. Whether we're subjected to a mysterious selection system or handed recommendations by a system we don't understand, our unconscious tendencies and the mysterious processes of technology and business are consistently coming together to change our behavior, put us in the loop, and make us into people we never consciously intended to be.

Everything we've discussed up to this point—the deep cognitive illusions and psychological vulnerabilities of the human mind, our tribalism, our persuadability, our deference to patterns—makes us the perfect data set for AI. Because as much as we like to believe that we are each unique, our behavior falls into broad patterns that it took more than a century of sophisticated psychological and sociological research to document, but that machine learning can spot in days, hours, seconds.

Aim a camera at a drunk couple at a baseball game and they'll kiss. Hold a hand up for a high five and a stranger will deliver. Lean too far back in a chair and the same thrill of adrenaline courses through everyone. Human behavior is patterns, and AI does nothing but spot those patterns and, increasingly, play to them.

As Weizenbaum might point out if he could, AI looks fancy, but it's no therapist. Machine learning samples only what has come before to make predictions about what comes next. And if it samples what we've discussed in the prior chapters—the fundamental programming of human beings, with all our biases, tribalism, allergy to uncertainty, and inability to grok big numbers—then our future will be built to reinforce those behaviors. When it comes to AI, the future is only an extrapolation of the past. Instead of making our lives better, machine learning may very well replicate and perpetuate all our worst flaws. In other words, if our unconscious tendencies are the objective function, we're in trouble.

And here's the thing: the most profitable AI is the kind that *does* sample those tendencies, because there's money to be made on who we are when we don't think about it. We're already geared to follow the instructions of System 1, our fast-thinking, instinctive brain, and all marketing is geared to sell to it. No company wants System 2 to come to the door. And for the same reason, no company wants to build AI that *empowers* System 2. AI that *compensates* for our ancient instincts will wipe out impulse purchases, reduce unthinking acceptance, kill off recommendation algorithms. No company wants that. The ancient instincts that AI is poised to spot and amplify—the tribalism, the anthropomorphism, the gut feelings and System 1 choices and irrational hatred of having our biases pointed out to us, all the stuff we started with in the beginning of this book—all that stuff is easy money once a company can teach AI to look for it. But how *does* AI look for it? And why do we believe AI is sophisticated enough to be ordering our lives for us? When we learn how it actually works, we can see why it *could* be used to empower System 2, but it is far more likely to be used to empower System 1. And that's where the third, outermost ring of the loop is taking form.

Chapter 7

WHAT AI ISN'T

EVEN THE PHRASE is trouble. From the first time humans paired the words "artificial" and "intelligence," we were playing with fire.

That first time was 1955, when a young assistant professor of mathematics at Dartmouth College, John McCarthy, tired of the disorganized overlap of nascent fields like "cybernetics" and "complex information processing" and "information theory," decided he wanted to bring colleagues together for the summer to try to create a little clarity around the concept of thinking machines. His funding application, which he coauthored with Harvard's Marvin Minsky, IBM's Nathaniel Rochester, and Bell Labs' Claude E. Shannon, was titled "A Proposal for the Dartmouth Summer Research Project on Artificial Intelligence." It's considered the first time the phrase was ever printed. The idea, as McCarthy and his coauthors put it, was to put a group together and "proceed on the basis of the conjecture that every aspect of learning or any other feature of intelligence can in principle be so precisely described that a machine can be made to simulate it."

Ten invitees attended some portion of eight weeks across the summer of 1956, paid for with $7,500 from the Rockefeller Foundation, and some four dozen people in all wandered in and out at some point, crashing in various empty apartments as well as the Hanover Inn, and generally freaking out together.

Thirty-year-old Ray Solomonoff, newly minted by the University of Chicago with a master's in physics, was the only attendee, aside from cohosts McCarthy and Minsky, to spend the whole summer at Dartmouth. He seems to have been the hype man of the group. He took extensive longhand notes on the gathering, embodying in his loopy logic and frenetic scribbles the communal frenzy of the experience. He even began to tinker with his personal food-and-sleep schedule so he could cram it all in.

"Try eating twice a day: noon and midnite [*sic*]," Solomonoff scrawls at one point, between staccato notes about predictions and optimum nonlinear filters. "Sleep 1am–9am. Only trouble is eating with others occasionally, but I think I could manage this. . . . The big disadvantage here may be not taking advantage of early waking acuity for problems worked on just before sleep."

This manic burst of creativity led to breakthrough notions, like the idea of a computer that can solve a problem it has never seen before by drawing on past experiences with other problems. It was a revolutionary idea. As Solomonoff wrote, a program "mite [*sic*] be given a series of correct examples, then a question without the correct answer." To make such a machine meant going against the traditional idea of a computer using only exactly the information presented to it. Perhaps, he wondered, ad hoc mathematical rules "might do only part of what I really need" and "non-ad-hoc rules are apt to extrapolate better." The ultimate difficulty, Solomonoff wrote, connecting half-finished paragraphs to one another with long, swooping arrows, was this:

> In any prediction problem we are confronted by a set of conditions from which we try to make a prediction. Ideally, this set of conditions will have occurred many times before, and accurate probability estimates can be made. More often, however, the set of conditions has never before occurred in exactly the same way. In such a case, we try to classify the event to be predicted, within a larger class of events, of which we have a sufficiently large sample, so that relyable [*sic*] probability estimates may be made.[1]

McCarthy and Minsky and Solomonoff and the various thinkers who floated through the gathering that summer wound up landing on a fundamental concept: a thinking machine that could take incomplete information and, using rules it had learned from past and parallel examples, predict that information's continuation, whether it be finishing a half-written paragraph of text or calculating the probability that a past event might occur again. The group was literally ahead of its time—it would be more than fifty years before what McCarthy and his crew described was technologically feasible at scale.

The gathering either launched or catalyzed epic careers for practically everyone who attended. McCarthy went on to develop an "advice taker," a computer program that used rules of logic (instead of just raw information) to solve a problem, as well as LISP, the major programming language of early AI. He developed notions of time-sharing and utility computing that gave rise to today's $250 billion cloud-computing industry. And he later founded and ran Stanford's AI lab, while Marvin Minsky ran MIT's. Wide-eyed, sleep-deprived Solomonoff went on to propose the first notions of "algorithmic probability" that could be used for predictions, and created the theoretical framework for using Bayesian statistics to deal with uncertainty, which makes him the ancestor of everything from modern weather prediction to AI that can spit out a reasonable-sounding term paper from a one-sentence prompt. RAND Corporation's Allen Newell went on to publish the first doctoral dissertation in AI, "Information Processing: A New Technique for the Behavioral Sciences." Prior to the Dartmouth summer, he and Herbert Simon (an economist and cognitive psychologist and seemingly the only Dartmouth attendee who wasn't a computer scientist) had already built Logic Theorist, the first AI program, and afterward they used military money to build the General Problem Solver in 1957, a program that simulated the rules of logic a human might follow.

But let's cool off from all this praise and excitement. Yes, the ideas were big, and the work it launched was revolutionary. But that's not the only reason this work set off the third loop, the cycle of behavior and pattern recognition and manipulation that may come to define the next phase of human life. It also set off the third loop because AI looks impossibly sophisticated and entirely inscrutable. We, the people whose behavior it will shape, just don't understand what it is and what it isn't. And so it's important to go back to the beginning of the field to understand the boundaries of what this group actually gave us. We need to trace the thread of their work forward from those feverish midnight conversations in 1956 to what we now call artificial intelligence today.

The full history of AI and its development has been documented by writers more thorough and more talented than I am, and I won't attempt to summarize the whole thing here, but it's worth understanding that the field fought with itself for years over whether computers could or even should be taught to handle open-ended tasks. Some of those objections were ethical, but mostly they had to do with market opportunity. IBM, where McCarthy went to work after the Dartmouth summer, was initially angling to be the corporate parent of artificial intelligence. But in 1960, a corporate study by the company detailed several reasons not to pursue that strategy, including surveys that found customers were hostile to—even afraid of—the notion of "thinking computers," and the company backed out. Minsky, Newell, and Simon were all prone to grandiose predictions in which robots would be picking up socks off the floor within a couple of decades. When that didn't bear out, the disappointment poisoned the reputation of AI. For nearly ten years beginning in 1974, the whole field's funding essentially dried up, the first so-called AI winter. And during the decades-long hunt for better systems, the field lost its way more than once. Minsky, for instance, published a whole book criticizing the idea of neural networks—interconnected systems of thinking "neurons" that might coordinate to attack a complex problem—and the field largely abandoned the concept for a decade. Neural networks are now a crucial part of modern AI.

And then there were the fundamental limitations of the era: a lack of computing power to crunch numbers, insufficient database capacity for the awesome amounts of information necessary to train an algorithm, and the fact that those limitations made it impossible to scale up small, one-off experiments into useful real-world systems. It took two funding winters, countless research dead ends, and exponentially greater computing power and data-storage capabilities to arrive at the present moment. Today, various flavors of machine learning, from deep-learning neural networks to the generative adversarial networks that pit two neural nets against one another, can do everything from read a printed menu to steer a car along a winding mountain road.

Again, this all sounds very hot—and it is amazing stuff. But what is actually being delivered into your life needs to be understood clearly, so we can see what it does and, more important, what it doesn't do.

First, artificial intelligence is not robotic intelligence poised to replace the many talents of the human mind. Before we get excited about that concept—a general artificial intelligence with the smooth-talking, all-purpose utility of Hasselhoff's K.I.T.T. car or Skynet—let's be clear that despite a famous survey of AI experts in 2013 showing that some 50 percent of them thought we'd have GAI between 2040 and 2050, (A) it doesn't exist at the moment, and (B) as we'll get into later in the book, all-purpose intelligence may not even have a market if simpler forms of AI take over our processes the way they likely will.

So, to pose a question that inspires even the most mild-mannered computer scientists to rain punches on each other, what is AI? Fundamentally, artificial intelligence is any system that can be assigned a task, learn from data, and adapt along the way. But there are, of course, many ways to accomplish this. I'll try to summarize a few flavors of it to give a sense of what we mean when we talk about decision technology, which will inevitably earn me a shouting at from experts at every event I attend from now on.

Machine learning refers to algorithms that get better at a task through experience. Machine learning draws on past patterns to make future predictions. But it cannot reach out beyond the data it has; to make new predictions, it needs new data. There are several forms of machine learning in common use at the moment.

First, *supervised learning* refers to systems shown enough labeled data and enough correct answers ("this is an orange; this is an orange that has gone bad; this is an orange that is ripe and healthy") that it can pick out patterns in the data. Ask it to identify specific outcomes (a ripe orange, an orange that will be ripe after a week of shipping, a rotten orange), and if it has seen enough of the patterns that correlate to those outcomes in the past, it can spot the patterns that will likely correlate to the same outcomes in the future.

Second, *unsupervised learning* refers to systems that are given data without any guidance—no right or wrong answers, no helpful labels, nothing—and attempt to sort them, typically into clusters, any way they can. Whatever common patterns differentiate the clusters become the sorting mechanism.

Third, *reinforcement learning* is another way of processing raw, unlabeled data, this time through reward and punishment. A training algorithm infers what you want out of the data, then flogs the system for incorrect answers and rewards it for correct answers. In this way, reinforcement learning teaches the system to sort out the most efficient means of avoiding punishment and earning praise. With no labeled data, it just goes on a craven search for whatever patterns are most likely to earn a reward.

Let's apply any or all of these three flavors of machine learning to a single task: distinguishing cows from dogs. Imagine we're in a theater. Ranged across the stage are a dozen dogs and a dozen cows. Some of the dogs are sitting, some are standing, but it's hot in the vast room, so all of them are panting. Among those dogs wander the poor, bewildered cows, complaining with anxious moos about the heat, the uncertain footing at the edge of the stage, the lack of grass and water. Now we'll ask our machine-learning system to tell us: which animals on this stage are cows, and which are dogs?

The trouble, as humans, is that we're going to want to believe whatever the system winds up telling us. System 2, our creative and rational mind, as we learned in the first chapters of this book, is only too happy to hand over decision-making to System 1, to our instincts, to our emotions, to any credible system that has a ready answer. And so it's important to prepare ourselves for just how tidily the algorithm will, eventually, deliver a reliable performance in differentiating the canines from the bovines. It's going to look extremely credible to System 1, and thus it will also look credible to System 2. And in a sense, it is credible. Feed photos or other data about what's on the stage into the system, and—*bing!*—out will pop a list, or captioned photos, of dogs and cows. With enough time to refine its process, the system will, eventually, astound us with its gift for telling us that this panting poodle is a dog, and that this heifer nosing the curtains is a cow. But how did the system get there? It matters because, as we'll see, this technology goes about answering stupid questions and world-changing questions the same way.

A data scientist handed this task would want to know more about the goal of the project in order to employ the most suitable flavor of machine learning to accomplish it. If you wanted the system to identify the cows and dogs in a photograph of the stage, for instance, you'd probably feed it into a convolutional neural network, a popular means of recognizing objects in a photograph these days. The system would, by the time you used it, have been trained on thousands of photographs of dogs and cows (supervised learning!), passing each photograph through a convolutional layer, which scans and simplifies each image, and then through a max pooling layer, which breaks the image into sections, and keeps only the section with the most representative information, before identifying the image as a dog or cow. (And in order to get good at the distinction, that supervised-learning process would have first been endlessly adjusted by human "trainers" who marked the results correct or incorrect, typically paid pennies per photo on a site like Amazon's Mechanical Turk.)

If, instead, the system was only given a list of attributes—size of animal, color, panting or not, hooved or soft-footed—reinforcement learning might simply begin guessing (dog! cow!) and adjust its verdict as the training algorithm compares those guesses to what it infers the humans want out of the matter. That process might take hours, days, maybe longer, but if no one had bothered to build a dog-versus-cow recognition system before, perhaps it would be the best route to distinguishing the animals from one another.

Whatever AI we brought in, there are two major forces at work that guide the outcome of the process. The first is called the *objective function*. It's what the human wants out of the project, whether that's a self-driving car parked equidistant between two other cars, no more than six inches from the curb, or a cheeseburger parked between two buns, cooked medium rare. The objective function is the purpose toward which the whole system is striving, and writing it out clearly is the defining first task of any successful machine-learning system. Ancient myths are full of badly composed objective functions: Pirithous wants Persephone, queen of the underworld, as his wife and winds up stuck in hell. Sibyl asks Apollo for as many years of life as grains of sand in her fist, but she forgets to ask for youth along with it and winds up shriveled in pain. If the self-driving car isn't told it must park *parallel* to the curb, or if the burger bot doesn't know the buns go on the outside, it can all go very wrong.

The other major force at work is the sheer ruthless efficiency that we design any machine-learning system to use in pursuit of the objective function. After all, the purpose of these systems is to save humans the time and effort required to spot patterns in a vast field of data, and to get better and better at the process over time. That means machine-learning systems working on the difference between cows and dogs are going to latch on to the first reliable means of differentiating the two, whether that's some combination of size and slobber (the dogs are smaller and, because they're panting in the heat, they're more slobbery; the cows are bigger, and while they're also suffering from the heat, they can sweat as well as pant) or the way the silhouette of the cows is more sharply defined against the curtain than the silhouette of the dogs, which happen to be fluffier.

Even if the two primary requirements are met—a clearly predefined objective function and a ruthlessness with which the system will pursue it—it will be hard, perhaps impossible, for you or me, waiting impatiently for our list of cows and dogs, to know that that's the case. In fact, it will be largely impossible for you and me to know anything about how the process works. Because although even if in the end our stage full of dogs and cows will eventually be categorized properly, the process by which our machine-learning system does the job will, by and large, be invisible to us. All we'll know is that we put an objective function into one end of the system, and after some trial and error we got a reliable-looking answer out of the other end. What happened in between, even for the specialists who *built* the system we used, is going to be largely a mystery.

As a result, there is, in fact, a movement in artificial intelligence referred to as "explainability," a push for transparency that is both ethically important and technologically difficult. Because while McCarthy and his crew would rightly be thrilled by what has been built since 1956, they surely didn't understand that extremely impactful decisions would be made by the sorts of automated systems they envisioned—who gets a loan, a job offer, the chance to get out on bail—without the ability to see *how*, exactly, those systems arrived at their conclusions. This is referred to as the "black box" problem. Modern machine learning gives us answers but doesn't show its work.

A 2018 competition called the Explainable Machine Learning Challenge, organized by Google, the Fair Isaac Corporation (FICO), and five universities, sought to force a solution. The FICO score is used in the United States to determine an individual's creditworthiness, based on an average score from the three main credit-reporting companies. (It is itself an infamous black box, in which Americans' ability to do crucial things like rent or buy a home is determined by a three-digit number into which they have almost no insight, and little recourse.) FICO provided the competitors with a data set of anonymized individual credit history information, along with whether or not each person had defaulted on a home loan. The challenge then asked the participants to build a black box system that would predict whether a loan applicant would go on to repay the debt, and to then explain how the black box arrived at its conclusions.

When deployed on big, unwieldy problems, every flavor of artificial intelligence tends to put together enormously complicated connections between the variables they are handed. For a human to identify how a system puts those variables together is itself as complicated as anything the system may have been designed to do. And that means that from the earliest prototypes, modern machine-learning systems have been more or less allowed to simply do what they do without much in the way of inspection. In fact, it's become a piece of accepted wisdom in machine learning over the years that if a system has to be forced to show its work along the way, it won't be as sophisticated or as accurate in what it can do.

This might not matter in small tasks like a cow-versus-dog test. But as machine learning begins to use human behavior as its data set, and predictions about what art we'll enjoy, what jobs we'll do well, what crimes we'll commit become its objective function, understanding the interstices of the system may be a matter of moral and legal necessity. At the moment it is not, largely because building a system that explains itself is simply harder to do than building one that doesn't.

Currently, building what's called an *interpretable* piece of AI involves observing certain constraints along the way. The designers might limit the number of variables under consideration so that they will know later that the system relied on at least one of a finite number of data points in reaching its conclusion. (The coat of the cows and dogs, their size, their shininess, their gait, but nothing other than those things, for instance.) The designers might combine smaller AI models that are themselves understood so that inspectors can decompose the system into its constituent parts down the line.

But it's not clear how one can reverse engineer the decision-making process of AI that hasn't been designed for transparency. The participants in the Explainable Machine Learning Challenge were being asked to build something intentionally opaque, feel around inside it for some sense of how it does what it does, and then try to turn that into a meaningful system of explanation. It was a horrifically difficult task.

And a six-person team at Duke University took a look at this Gordian knot and decided to cut right through it. They looked at the FICO data and the requirements of the competition and decided they weren't going to build a black box at all. They believed they could build a system that didn't have to be a black box. It would issue accurate verdicts the average banking customer could inspect and understand.

Professor Cynthia Rudin, a computer scientist on the Duke team, and Joanna Radin, a sociologist and historian at Yale, wrote about the team's experience in an article for *Harvard Data Science Review*:

> The model was decomposable into different mini-models, where each one could be understood on its own. We also created an additional interactive online visualization tool for lenders and individuals. Playing with the credit history factors on our website would allow people to understand which factors were important for loan application decisions. No black box at all. We knew we probably would not win the competition that way, but there was a bigger point that we needed to make.[2]

And they were right: they didn't win. The organizers of the competition would not allow the judges to play with and evaluate the Duke team's visualization tool, and a trio from IBM Research won the competition for a system based on Boolean rules that would help a human data scientist inspect the black box. In its paper describing the winning system, the IBM team rightly pointed out the need for explainability "as machine learning pushes further into domains such as medicine, criminal justice, and business, where such models complement human decision-makers and decisions can have major consequences on human lives." That said, their black box would have been just as useful to FICO whether it was later explained or not. Nothing about its design *required* FICO or you and me to understand its inner workings. It just made doing so an option.

But FICO, to its credit, looked at the Duke team's highly accurate and user-friendly system, decided that project was worth rewarding, too, and invented a prize for it for going "above and beyond."

The problem is, black box systems are already a mainstay of modern business. Executives who depend on AI are just as oblivious to the mechanism at work under the hood as you and I would be to how our algorithm told those cows and dogs apart. FICO went on to conduct a 2021 survey of C-suite financial-services executives from companies that make more than $100 million in annual revenue. It found that nearly 70 percent of them could not explain how their AI systems made decisions or predicted outcomes. And what's more, by and large they didn't seem to care. Only 35 percent of surveyed executives said they'd made any effort to bring accountability or transparency to the process.

Black boxes are, of course, just easier to build. Wind it up, let it go, don't worry about it. But they may have become the norm for a few alarming additional reasons. First, there's the assumption that letting a system learn its own rules behind closed doors is the fastest path to an accurate result. Then there's the desire to keep the inner workings of a valuable system from getting out to competitors, even if that means those who built the system don't understand the inner workings, either.

However, there's also the fact that demonstrating cause and effect is the first step in any lawsuit against a company deploying a technology like AI. Based on a 1975 federal case, *Warth v. Seldin*, it has been necessary for plaintiffs to establish clear causation linking their loss or injury to a defendant for a lawsuit to go forward. But if plaintiffs can't even inspect an AI that has denied them a bank loan, or a job, or a shot at bail, how could they even begin to make that claim? "To some, the obvious solution to the intent and causation problems," wrote the attorney Yavar Bathaee in a 2018 article in the *Harvard Journal of Law & Technology*, "will be to either increase the transparency of AI through standards-based regulation or to impose strict liability." And yet, he argues, "both approaches would risk stifling innovation in AI and erecting steep barriers to entry." Whether he's correct on the effect regulation might have, it's clear that a company that builds a black box system is currently dodging a whole host of legal and public-relations headaches. It may be that we've come to consider black boxes not only more accurate and easier to protect from competitors, but that we've also adopted them as a bulwark against legal exposure.

The black box paradigm "allows the model creators to profit without considering harmful consequences to the affected individuals," Rudin and Radin wrote. "It is possible that an interpretable model can always be constructed—we just have not been trying."

There are companies that have tried. Verikai is an AI-driven risk-assessment company that tries to give insurers the ability, through machine learning, to underwrite health insurance for companies smaller than five hundred people. Doing so has been very difficult in the past, because the law of big numbers in health insurance meant needing the largest possible group of employees in order to accurately predict how many of them in a given year might develop diabetes, cancer, or other diseases. Below five hundred people the predictive power begins to fade, and below two hundred fifty people it tends to fall apart completely. Verikai has instead sought to extrapolate risk by using AI to spot patterns in enormous numbers of people, and then carrying those predictions into small pools of employees. Using data brokers like Acxiom and Experian, among many other sources, "we are scoring people's sleep patterns, alcohol consumption patterns, eating patterns," Verikai's former president, Chris Chen, told me. AI is essentially allowing Verikai to draw conclusions about what humans do *in general*, and then carry those conclusions into smaller representative groups.

People who report better amounts of sleep, or who consistently adhere to taking their medication, correlate to a better chance they'll repay their mortgages, Verikai has found. But "there's other really random correlations," Chen told me. And this is where AI is not only making deeply influential recommendations—recommendations that will affect fundamental aspects of your life, like whether your employer decides it can afford to help you pay for health insurance—without showing its work. It's also making deeply influential recommendations based on stuff you didn't even realize it was using to reach conclusions about you. "You own a swimming pool and you have, say, these three other attributes and as a result, you're lower risk and we're not sure why," Chen offered as an example. But he also doesn't believe companies need to blindly accept those correlations. "Everyone talks about AI needing to be a black box," he said. "It really doesn't need to be." Instead, Verikai's system is built to actually show the public what aspects of their lives suggest a certain degree of health risk, according to the patterns the AI has spotted.

The company's AI draws correlations between seemingly unrelated predictors of your health outcomes, but at the very least the company can show you what they are. Most won't do you that courtesy, and there's little incentive for them to start. By and large, we have no idea how these technologies are working on us; we just assume they work.

FOR ROUGHLY THREE hundred years, one irrational theory dominated the thinking of rational people trying to explain the universe. It was a theory that sprang from ancient superstitions, and for which there was no measurable evidence, but it was so useful in its simplicity that it became an article of faith to which generations of researchers remained committed even into the twentieth century.

It was the idea of "the aether," an absolute stillness in the universe against which movement—like that of light—can be measured.

Plato and Aristotle got it started by referring to a "fifth element" (*quinta essentia*) that sat in some way above Earth, Air, Fire, and Water. (The ridiculous and wonderful 1997 film *The Fifth Element* revolves around the idea that the *quinta essentia* is love, but that's not what we're talking about here.) The philosophers each theorized that this quintessence made up everything beyond our sky. And while most ancient Greek ideas about it were soon dismissed—that it moved in a circular pattern, for instance, while the other elements somehow traveled along linear paths—the idea stuck around that there was a "pure" element in the universe, a clear and untainted medium through which things move.

Early medieval alchemists were fixated on the idea that all things had a quintessence—a purest form—and that through the proper chemical processes they could distill it from various substances and use it as a curative. The fourteenth-century alchemist John of Rupescissa popularized the idea that an alcohol-based quintessence could heal affliction and forestall death (an attractive notion at a time when most people, including Rupescissa, died before the age of fifty).[3] Alchemists tended to defend their work with mystical explanations, however, and by the end of the medieval era, as the Renaissance ushered in a new era of evidence-based science, the notion of quintessence as a magical medication had largely faded away.

But the *idea* of a pure essence lived on, and in the seventeenth century René Descartes began to formalize it in the realm of physics. He was trying to move beyond what he considered outdated medieval theories when he used the term "aether" to describe a great ocean in which everything swims: in his view, the aether was a theoretical medium across which distant objects were still physically connected, and through which something like magnetism "moves." This notion helped to create a mechanical explanation for the behavior of light, and physicists continued to depend on and refine it. While Descartes considered the aether to have a static property, Robert Hooke theorized that it vibrated, Christiaan Huygens theorized that it consisted of spinning particles, and then Isaac Newton wiped them both out with his suggestion, delivered in the same memorandum to the Royal Society in which he explained gravity, that "all space is occupied by an elastic medium, or *aether*, which is capable of propagating vibrations in the same way that air propagates the vibrations of sound, but with far greater velocity."[4] In Newton's view, small, rapidly moving corpuscles made up the aether. His theory stood for at least a century.

Newton's idea lasted so long perhaps because it was so handy, a point of reference that made calculations easier and theories more convincing. And by the nineteenth century, the existence of an aether was not only assumed to be a fact, it had been built into other theories about the sprawling mechanisms of nature. An 1889 correspondence between two pioneers of electromagnetism, Oliver Heaviside and Heinrich Rudolf Hertz, shows just how deep this faith ran. "We know that there is an ether," Heaviside wrote to Hertz with a certainty typical of the era, while acknowledging with frustration that it had been unobservable. "As for the structure of the ether itself, that is a far more difficult matter, and one that can never, it seems to me, be answered otherwise than speculatively."[5]

But then a new generation of nineteenth-century physicists began to squint at the whole notion of the aether, because certain aspects of the theory just weren't working out. For one thing, it wasn't clear why something like our planet could move through space without any obvious sign of being slowed by the friction an aether should theoretically impose. Instead, the Earth travels around the sun at a steady 67,000 miles per hour. And if it were moving across an aether, that aether should give off some sort of wind, and Earth's movement should be affected depending on whether the Earth was traveling with that wind or against it. And as these granular questions became a priority, cutting-edge experiments began to reveal that generations of scientists had been entirely, embarrassingly wrong.

Between April and July 1887, a pair of American physicists, Albert Michelson and Edward Morley, conducted experiments on the outskirts of Cleveland at what is now Case Western Reserve University. Michelson had prototyped the interferometer they'd be using for the work while doing research for the navy, and spent most of the prior few years laboring so relentlessly toward the goal of detecting aether wind that he'd had a nervous breakdown in 1885. When he and Morley finally settled into their spring and summer together in 1887, safe from interference in the basement of a stone dormitory, they had very high hopes they'd built a device sensitive enough to detect the difference between light traveling with the aether wind, and light traveling against it. And they had. But when their interferometer split light from an oil lamp, sent each beam at perpendicular angles up and down the arms of the device, and then brought them back together again, the beams arrived at the same time. There was none of the delay they'd expected. "The Experiments on the relative motion of the earth and ether have been completed and the result decidedly negative," Michelson wrote to a British counterpart. And in experiments by other physicists in 1902, 1903, and 1904, the same results came back. Suddenly it seemed that the all-important and all-convenient aether was not a thing. It didn't exist.

With the idea of a constant medium no longer workable, old theories collapsed, new theories were necessary, and a physicist born in Germany just a few years before Michelson and Morley emerged disappointed from their basement lab came up with the best theory around. In 1931, long after he'd achieved fame for his scientific achievements, Albert Einstein met Albert Michelson for the first and only time at a dinner in Einstein's honor at Caltech. According to his biographer Albrecht Fölsing, Einstein gave an after-dinner speech for a crowd of two hundred that included the aging physicist, and thanked Michelson directly for the work he had done when Einstein "was still a little boy, hardly three feet high."

"It was you," Einstein said to Michelson, who died just four months later, "who led physicists onto new paths and by your wonderful experimental work even then prepared the road for the development of the relativity theory."[6]

What Albert Einstein established from the ashes of the aether was that the universe was a whole lot more complicated than Plato, Aristotle, Descartes, Newton, or anyone else had understood. And yet he showed we *could* measure it. We just couldn't measure all aspects of it at the same time. His theory of relativity had shown, for instance, that space and time couldn't be measured separately from one another—they were connected. And because of the space-time continuum, and the effect that massive objects have on it, events that take place at the same time for you can take place at different times for me. In fact, Einstein had found, to gather the information necessary to identify one fact about an object—like where it's headed in space—an observer would have to sacrifice the information necessary to identify another fact—like its position at that particular moment. There was no stillness. There was no ocean across which things moved and against which they could be measured. Everything is moving in relation to everything else all the time, and our math had to adapt to this new fluid model.

I believe we are still in a phase of decision technology in which we naively consider there to be some sort of simple aether against which we can measure the effectiveness of what we've built. And it's called an objective function.

In a post on the popular AI blog *Abacus*, Daniel Kronovet describes the notion of an objective function in a way that Descartes, Newton, and other pre-relativity physicists would have recognized:

> Put metaphorically, we can think of the model parameters as a ship in the sea. The goal of the algorithm designer is to navigate the space of possible values as efficiently as possible to guide the model to the optimal location.
>
> For some models, the navigation is very precise. We can imagine this as a boat on a clear night, navigating by stars. For others yet, the ship is stuck in a fog, able to make small jumps without reference to a greater plan.[7]

But AI, especially when it's called upon to parse the patterns in human behavior, is navigating something much more complicated than water, toward something much more complicated than the closest destination, and the objective function for you may be very different from the objective function for me.

I once sat in a meeting where a group of researchers put their prototype for a piece of "ethical AI"—a system one could trust to take ethics into account as it did its work—in front of a roomful of behavioral and political scientists. (I attended under an agreement that I could describe the meeting's content, but not its participants.) The lead researcher, clicking through the slides behind him, described his team's plan: Build a long list of half sentences along the lines of "As a coworker, it would never be appropriate to [blank]." Have humans complete a couple thousand of those sentences. Then train AI to spot whatever hidden patterns there are in the result, and begin filling in the blanks itself. Given enough training data, he concluded, putting the remote down on the desk, "we think we can arrive at a set of universal human values. Thank you. And now I'll take your questions."

Every hand in the room went up. A political scientist went first. "I guess I have three questions for you," she said. "First, what is 'universal'? Second, what is 'human'? And three, what are 'values'?" And with that the meeting imploded.

Social scientists, without using the term, have been squinting at whether our species has an "objective function" for more than a century. Kahneman and Tversky and Thaler and Banaji and all of them have looked at the fundamental programming that makes us who we are, and for the adjustments that can be made to make us better. But increasingly, a new crop of social and political scientists are beginning to step back from the problem and ask whether in some cases—or even most cases—there's any universal notion of "better" we could ever agree on. They're pointing out that *better* for you or me could be in fact far *worse* for someone else in other circumstances. And what they're pursuing suggests that not only isn't there a sea across which we're trying to point a ship, or a destination to which we can navigate, but also that it's in fact *dangerous* to build an automated system that measures success against a universal medium, in service of a universal goal.

At this point it's probably clear that I'm laying a great deal of my concerns at the foot of AI itself—its craven devotion to pattern recognition, our overinflated opinion of it, the way it hides its work. But I don't believe we'll be able to protect ourselves from the third loop by only adjusting the way AI is built. We also need to recognize that the patterns in our behavior don't always point the way toward a desirable objective function. Using AI to make important choices in society isn't just problematic because of the AI. It's also problematic because some of our choices are impossible.

John Patty and Elizabeth Penn taught me this. They're a husband-and-wife academic team at Emory University. They're both professors of political science, but they're also mathematicians, and while other political scientists tend to work in formal theories that treat humans as rational actors with rational goals, these two use math to try to account for the persistent inability of humans to get along. In effect, while others are still trying to measure the physics of human choice against the aether, Patty and Penn have come to accept, just as Einstein did, that things are much harder to measure than we thought they were.

In their 2014 book, *Social Choice and Legitimacy*, Patty and Penn write about whether we can ever bring conflicting individual goals together into a grand collective decision. And almost immediately it's clear that these two are trying to account for the thorniest possible stuff. Because rather than suggest it's possible, as civics class always tried to teach me, that people can come to a satisfying consensus through earnest discussion in a town hall or a national election, Patty and Penn suggest that satisfaction isn't the thing we should be after.

"In a nutshell," they write, "any minimally democratic aggregation procedure must encounter some situations in which it fails to produce a coherent (or, perhaps, 'well-ordered') collective preference." As a group, there are some things on which we simply cannot agree. So what can we hope for? Patty and Penn argue that rather than just give up on the idea of a popular will, as so many of their peers have spent careers arguing, we should instead hope to establish a feeling of legitimacy in how decisions are made, by explaining them and the principles they're based on. They know that won't be satisfying for everyone. It won't make for a great speech at the end of the movie. It won't create a catchy tune. "Our theory does not provide any particular reason to expect that all individuals will be equally happy (or, perhaps, happy at all) with the conclusions reached by such a process. But that is arguably the point of legitimacy: a legitimate decision is one that makes sense in spite of not being universally acclaimed."[8]

They aren't writing about AI or objective functions, but they might as well be. Their theory of social choice, which one reviewer called "a thicker conception of democracy,"[9] suggests that in order to achieve any satisfactory outcomes in law, politics, or administration, we must feel that the legal, political, and administrative system knows what it's doing. And we have to have the principles that drove a decision explained to us. None of that will be possible if we continue to go down the road of deploying opaque, proprietary AI systems to evaluate our world and render verdicts for us.

The expediency of business isn't just causing companies to ignore the inner workings and ethical risks of the AI that makes their services possible. In fact, most companies are reusing one piece of AI over and over again. It is prohibitively expensive for most companies to purpose-build AI for the task at hand. Instead, for reasons of cost, they adapt existing algorithms, built around other objective functions, for tasks in some other domain. And as that has become more and more the case, the same pieces of machine learning are being deployed on everything. This creates the risk that certain reused pieces of AI are going to be applied to tasks for which they aren't designed, and may carry over algorithmic habits learned in other data sets that wind up skewing important results in the one to which it's been ported. As UC Berkeley computer science professor Moritz

Hardt put it to me, when it comes to developing algorithms that are fair and effective, "lending is going to be different from criminal justice, and college admissions are not the same as recidivism prediction." And when it comes to getting things wrong, "the cost to society, to the individual, and to the decision-maker, they're very different." Bolting the same forms of AI onto each problem can save and make money. But those objective functions shouldn't be considered universal, he told me. "There's not going to be an easy, all-purpose definition that we could just sort of slap onto our algorithms and solve the problem." And yet in a world where so few leaders of companies care to even inspect *how* the AI they use works, because the technology seems to deliver reliable, efficient results, who will be on guard against subtle, hidden algorithmic effects and biases that don't show up in the bottom line, and may take years to play out?

I wish it were the case that the solution to this problem is simply improving the algorithms, or requiring explainability, or demanding that companies tell us when they've borrowed AI trained on recipe recommendations to build a system that sorts through résumés. But it's not just a design problem.

Once upon a time this sort of world-changing technology was developed by academic institutions, national labs, the Department of Defense. Today, AI is being refined entirely inside for-profit companies. It's built to make money. And yet the marketing, the libertarian politics, and the continued social acceptability of technologically driven industries like the one that has produced decision-guidance technology depends on the assumption that tech makes our lives better. That assumption drives the adoption of AI, as it drove the adoption of smartphones and the Internet before that. It's a reflection of our fundamental Western assumption that things are always improving, and of our fundamental American assumption that capitalism creates solutions that will lift up everyone. I worry that in AI we've built something that will so fundamentally reshape human behavior in pursuit of profit, and will so easily find and amplify its worst aspects along the way, that all of those assumptions will be proved wrong without our realizing it.

The Unitarian minister Theodore Parker Jr., speaking in 1853 about the prospects of abolishing slavery, said, "I do not pretend to understand the moral universe; the arc is a long one, my eye reaches but little ways; I cannot calculate the curve and complete the figure by the experience of sight; I can divine it by conscience. And from what I see I am sure it bends towards justice." In 1950, at the end of the Montgomery bus boycott, Martin Luther King Jr. condensed the idea in a speech about the prospect of equality for all Americans. "How long? Not long, because the arc of the moral universe is long, but it bends toward justice." President Obama often cited King's phrasing in his appearances at the podium.

It's a beautiful idea, and it may yet prove to be true. But consider that we have only been thoughtful enough to express that sort of idealistic sentiment for a few thousand years, and to have ideas like that at all for perhaps 70,000. The history of humanity simply isn't long enough to know whether it's an arc, or which way it bends, as Parker originally admitted. We're not on a preordained flight path toward a better life. It's not a given that what we build will improve our world. We are living an improvisation, carefully balancing our institutions, our partnerships, our social codes on the tip of a collective finger we only just figured out how to use. What is universal? What is human? What are values?

IMAGINE TRYING TO explain the procedure of eating at a restaurant to someone who has never been inside one.

Well, first you'll walk in, and there will be tables everywhere. You may or may not be able to see the kitchen. Regardless, don't sit in the kitchen. Sit at a table. But don't sit at a table unless it's obviously clean, and has a knife and fork set out for you. Although some restaurants don't do that—the table might be blank. Also, wait at the entrance for a few minutes to see if someone at the restaurant wants you to sit somewhere in particular. Okay, now sit down.

Mark Riedl has spent years trying to do this maddening task. Not just writing a guide for proper restaurant etiquette. He's trying to teach all forms of human behavior to robots. But it's nightmarishly complex. Every rule of human interaction has a subtle exception, and the branches of the decision tree bloom endlessly. "At some restaurants, we wait in line," Riedl says. "But in other cultures, the rules about waiting in line are different, and people cut the line."

There's no manual of human interaction, Riedl sighs. "If you want to learn the rules of society, then you have to go learn it somewhere."

Riedl began his academic career in 2008 trying to create computer games that could rewrite their own plots as the story progressed. "I had an advisor at North Carolina State who was trying to use AI to manipulate computer games. He wanted to get out of the conventional plot, like if you wanted to suddenly join the bad guys. But to do that you had to build a story generator." And that's where a lot of research had dead-ended. Computers could string together prewritten plot points in random order, but they didn't know how to assemble a plot that would resonate with our human expectations.

So Riedl turned to a singularly human source of knowledge: our stories. "Our ability to pick up a story and absorb its lessons is one of our great talents," he says. And he began working on the idea of using stories to teach robots what to expect from human society.

Unfortunately, that turns out to be very hard. "For one of my story-understanding systems," Riedl says, "we did the math, and the branching choices number in the thousands." To a computer, ordering food at a counter is outrageously complex, even though to a human, Riedl says, "a restaurant is a small, rigid, agreed-upon human interaction."

In 2016, Riedl and his colleague Brent Harrison, working together at the Georgia Institute of Technology, published a paper, "Learning from Stories: Using Crowdsourced Narratives to Train Virtual Agents." In it, they introduced Quixote, a piece of software that could absorb stories from humans and distill teachable rules from them. For one of its first tests, they tried to teach Quixote the rough rules that govern a bank robbery.

Their heist scenario involved three characters: a bank teller, a robber, and a police officer. Each of them had only a handful of available actions, from pushing the alarm button to brandishing a gun, and yet the branching combination of choices and resulting story lines for Quixote as the director of the scene numbered in the millions. At first Quixote had no idea how to write the script of the holdup. After all, it has no inherent interest in plot or tension, and no instincts for what we'd expect from the scene. It's software. It wants to get the thing done as efficiently as possible. So it just strung plot points together for the easiest possible transaction. The cop would simply stand aside so the robber could leave unhindered, for instance, or the robber might take the money and then hang around the lobby endlessly, waiting for the cops. So Harrison and Riedl recruited people online to write short descriptions of a typical bank robbery in simple English that Quixote could read.

Soon Quixote was scripting bank robberies that a Hollywood screenwriter would recognize: the thief pulls a gun, the teller hits the alarm, here come the cops, the chase is on. It worked. And in the process, it proved Riedl and Harrison's concept: that telling stories to robots could teach them what humans typically do.

Their work is largely funded by the military, either the Office of Naval Research or DARPA, both of which are deeply invested in the idea of robotic helpers. The military as a whole has committed billions of dollars to these sorts of projects.

The grant officers from ONR explained their interest in Riedl's work two ways. First, the military wants to make it easier for unskilled humans to program a simulator. "Let's say it wants to build a social simulation of a town in a foreign culture," Riedl says, "but all the subject-matter experts are not programmers. How do you teach the program the details of the place, the characters, the farmers in the field, what they'll all do when you show up?"

The other military interest is building robot systems that will get along with the humans operating them or operating alongside them. "We tend to assume the robot will act like a human, and when it doesn't, we're surprised," Riedl explains. In a future where a robot is carrying soldiers' belongings, or driving them around, the soldiers need to be able to reasonably anticipate what the robot will do next. And if they can't, our human tendency to anthropomorphize anything that looks vaguely human gets us into trouble. We shouldn't expect the robots to behave in human ways, and yet we do. Riedl says that one of his goals in telling stories to robots is to keep them from behaving in ways we find off-putting—he uses the word "psychotic." Telling human stories to robots, Riedl and Harrison hope, can perhaps help them act more human.

At this point we've talked about several forces converging in the field of AI: the way the technology makes decisions by drawing correlations between factors we ourselves don't understand, the opacity of that process, our eagerness to build businesses on the incredible efficiency that AI offers, our historical tendency to assume things are simpler than they are, and the growing realization that human decisions are more predictable and human goals more complicated than we might like. And now here's an added problem faced by Riedl and Harrison and everyone else trying to teach human decision-making to robots: not only are most of our decisions unconscious, and not only are they often rooted in instincts we'd consciously reject if they were pointed out to us, human society is built to sort out the rightness of our actions only *after* we've made our choices.

If a child suddenly leaps into the path of an oncoming truck, the truck driver is faced with an instantaneous and impossible set of branching decisions. Plow ahead, presumably killing the child? Pull the wheel one direction, steering the truck into oncoming traffic, where countless others could be maimed or killed? Or pull the wheel the other direction, taking the truck off a cliff?

Once that driver has made the decision and the horror has ended, the ambulances have carried away the injured and dead, and the police report has been filed, somcone has to evaluate the choice he made.

"It was instinct," the driver keeps saying. "I don't know why I decided to do what I did." So the investigation looks at questions of sobriety, at the mechanical integrity of the truck, at the parents of the child. And at the end of it all, the wheels of justice—the police, the courts, the juries—arrive at some sort of conclusion that determines fault and restitution and the policies that might prevent this sort of tragedy from taking place again.

With autonomous systems, that kind of evaluation has to take place *ahead of time*. An autonomous vehicle will only do what its rules tell it to do. It must be trained in advance to properly evaluate whether to go ahead and plow through a chicken in the road, but veer away from a child. We must teach it to choose the cliff or oncoming traffic.

Can enough stories teach a robotic system to make a perfect decision every time? "We don't write our values in a logical, coherent way," Riedl says, echoing what that roomful of social scientists all had their hands raised to say about human values, echoing what Einstein helped to establish about the relativity of time and space, echoing what Patty and Penn have to say about giving up on consensus and settling for legitimacy. This is the great challenge of "ethical AI." Riedl and Harrison and everyone else are trying to use pattern recognition to pick out a coherent set of human values when we humans don't yet understand them ourselves. "We're asking autonomous systems to be perfect, and yet we tolerate errors in humans," Riedl says. He thinks for a moment. "I don't have an answer for that."

AT THIS POINT it's hopefully clear that we're faced with a conflict between AI, our ambitions for it, and our tendencies and needs as individuals and as a society. We give AI an objective function and expect it to reach that goal with ruthless efficiency—we want an aether against which AI can be measured—but setting objective functions for humans, given how different we want to be from who we naturally are, is inherently problematic. But let's assume, for instance, that the function of AI should be to make life better for humans in general. How would we even begin to determine what the objective function of that algorithm should be?

One AI researcher told me in 2019 that he and his colleagues had been kicking around a hypothetical ethical dilemma they called "the heroin problem." (It's important to recognize here that the hard, yearslong work that harm-reduction experts have done to reframe opioid abuse as an affliction, rather than a choice, makes the binary example he laid out a dated and insensitive one. So also consider that AI researchers are kicking these ideas around, but that they're not often up to date on the latest thinking outside their field.)

He described the heroin problem to me this way: Imagine being asked to build a deeply personalized AI-based assistant, one that looks for patterns and automatically offers proactive help, in order to help the customer reach her goals as efficiently as possible. Let's imagine that most users adopt the product in order to help save money for retirement, and as a result the AI begins looking for signals in the data that correlate to financial growth. The objective function, in other words, is "save as much money as possible." (This is not yet a thing, it should be noted. No one makes money tricking people into not spending theirs. It's the reason you've never seen a flashy television advertisement for a Roth IRA or a 401(k).) The AI might instantaneously sock away any extra income in an investment account, and politely offer to cancel appointments that have been linked to spending money in the past.

"But now imagine you're addicted to heroin," he said to me. "What should the objective function be?" Should the AI draw its lessons from all the other people it has analyzed, people who are saving money, and try to nudge me off the habit? Or should it accommodate my dependence on heroin, by, say, prioritizing time with other people who use, or shielding me from appointments that past patterns suggest I'm going to be too sick with withdrawal to attend?

And even if AI could be entrusted to pursue these sorts of goals, how do we know that the people making it are even interested in pursuing a form of the technology that can handle the variety and subtlety of human behavior and satisfaction? AI is being developed in pursuit of profit, not perfection, after all. As a result, it's not clear that the handful of companies that dominate the field have any reason to squint skeptically at the short-term conveniences of AI the way that nineteenth-century physicists squinted at the short-term conveniences of the aether theory. Imagine that while Michelson and Morley tuned the arms of their interferometer in the cool stone basement, researchers elsewhere around the world had discovered that one could make vast sums of money off aether-based calculations, whatever the abstract, long-term threats to human understanding of the universe. If the aether theory had promised to save endless hours of calculation, and produced results that *looked* reasonable and produced a return the way that AI has disrupted industry after industry by matching advertisers to potential buyers and political opinions to an audience, we'd likely still be talking about aether winds and a theoretical universal substance. The uses to which AI is being put aren't about truth and science and deeper understanding—at least, that's not why companies are putting money into developing the technology. Something like the "heroin problem" is difficult enough on its own, but it foretells disaster when you consider the profit-making purpose of AI.

And so although it should be clear by this point that using AI to make money could unleash a decision-making force in society that we'd have no control over, leading us blindly in all sorts of terrible directions, none of that gets in the way of the business case for using AI. And that's where The Loop has begun to encircle us.

Chapter 8

COLLAPSING SPIRAL

THE FINAL RING of The Loop—the one for which this book is named, the convergence of pattern-recognition technology and unconscious human behavior—is still in disparate pieces. It hasn't fully taken form. But its component parts are being assembled all over the world, in all sorts of industries, and the way they can be bolted on to any part of your life (an algorithm that can pick out patterns in your taste in wine can also credibly pick out patterns in the clothes you might wear or the live concerts you might enjoy) means they're going to naturally click together over time. And if that happens before the behavioral effects become grounds for legal action or regulation, we'll see a world in which our choices are narrowed, human agency is limited, and our worst unconscious impulses dominate society. Some people seem to believe that AI will free us, that we'll wheel through the world unencumbered by busywork or uncertainty. I worry that as we become caught in a cycle of sampled behavioral data and recommendation, we will be instead caught in a collapsing spiral of choice, at the bottom of which we no longer know what we like or how to make choices or how to speak to one another.

I want to explain in this chapter how easily decision technology spreads into all corners of our lives. And while the examples I'm about to describe may feel disconnected, remember that the interoperability of machine learning means a set of algorithms built to do one thing can also do many others well enough that you'll never know its various roles, so anything AI can do in one part of your life will inevitably metastasize into others. I could start describing this third ring of The Loop with examples from all sorts of industries. But I want to start with surveillance, because it's immediately understandable, it's already changing our behavior in unconscious ways, and its effects are rapidly accelerating.

After five weeks of lockdown at the beginning of the pandemic, when it first became clear that this would be a deep and vast economic crisis, with more than 26 million Americans filing for unemployment by late April, the national and international mood turned toward a desire for some way to monitor the spread of the virus while somehow getting back to public life. The president of the Veneto region in Italy, where the first known Italians had died of the virus, talked about creating some sort of work-license system for people who had tested positive for COVID antibodies. Public health officials in the United Kingdom and elsewhere began openly discussing the possibility of a "passport" given to those with the necessary antibody test results or a vaccination card that would allow them to work and travel again. Google and Apple announced a plan to build an app together that would use Bluetooth to detect other phones nearby, and, if a person volunteered to have their test results and contacts recorded, it would alert users if they'd been too close to someone who was later determined to be infectious. The logic of building systems of detection and certification was obvious. And determining who was no longer a risk to the public seemed to require some sort of central public database. The involvement of ubiquitous companies like Google and Apple seemed inevitable and inarguable. Meanwhile, I began receiving email after email from companies that claimed to have built contact-tracing systems, because technologically they were relatively simple to build: any company with a location-aware app for following the behavior and location and social connection of its users already had, in theory, the constituent parts to build a contact-tracing system. This was a rare moment in which building surveillance technology was a matter of public interest and public health.

Soon a new set of radical surveillance capabilities, largely powered by AI, became public knowledge. A company called Draganfly, which had invented the original four-rotor "quadcopter" design for drones, contacted me to describe a new technology that it said would enable public-health authorities to detect COVID-19 symptoms from the air.

Researchers at the University of South Australia had already created technology to rapidly detect, in a single drone pass, the survivors of a mass-casualty event like an earthquake or tsunami. With 4K-resolution video footage held extremely still by stabilizers, AI could compare the images in the footage to training data showing the motion of humans breathing in and out. Once the system had seen enough footage of people breathing, the researchers found, it could pick out the living from the dead according to the tiny movements of even the shallowest, desperate respirations.

Now, they had found that other vital statistics could be gathered in a similar way. Not only could the system learn to spot breathing, it could subdivide the rate and depth of the breaths and correlate it to unhealthy types of increased respiration, and to elevated heart rate. The movement of people coughing into their elbows could be classified according to the severity of the cough. And the technology could analyze a person's skin tone and compare it with people of similar skin tone to spot people who were especially flushed. And by putting skin tone and heart rate and respiration together, perhaps the technology could even spot a fever. Maybe this was the makings of a respiratory-virus detection system.

The researchers had been flying their technology on one of Draganfly's drones and contacted the company about a deeper partnership. By April 2020, the company was ready to fly its expanded vital-statistics program wherever they could find permission to do so. And in Westport, Connecticut, the police department said yes.

The Westport Police Department already operates a technology-heavy surveillance platform, as its chief, Foti Koskinas, explained to me. He's a rapid talker and takes obvious pride in his work. Many of Westport's 28,000 residents commute into Manhattan, and several financial-services firms have offices in town, making it a busy train stop. What the roughly 6,000 people getting on and off a train at the station may not realize is that the police scan their faces as soon as they hit the platform. "The police department has almost 200 cameras that we monitor through our train stations and other areas that fall under our jurisdiction," he said. "We do have facial recognition on those cameras, we have plate readers. So we're doing many things to gather data." Westport is one of the Eastern Seaboard's most prosperous and homogeneous towns, with a median household income of nearly $200,000 and a population that's more than 92 percent white. The accusations of racial bias that have been leveled for years at facial-recognition systems didn't seem to have triggered much hesitation on the part of the police department or city government when it came to deploying it on the citizens there.

Furthermore, Chief Koskinas said his department has regularly flown four drones over the town since 2016. "We have so many parks, so many school campuses, beachfront, an island, we have so many areas that my patrol shifts cannot possibly adequately cover all of it," he told me. The drones have been used to inspect roofs, find missing people, and were already being used to detect and disperse crowds of people identified by Draganfly's software as standing too close to one another in the age of social distancing, when the company proposed using Westport as a test bed for its symptom-detection software.

As the CEO of Draganfly, Cameron Chell, described it, the purpose was not to identify any one person in particular. Instead, the first phase of the technology would be geared toward generating a sense of the rate of infection in a given area.

But soon, Chell told me, he believed the nation would need ongoing monitoring to get the economy up and running again. "The consensus that we seem to be getting back is that the new norm is health monitoring on a broad basis," he said. Perhaps to reactivate something like sports or professional gatherings, a drone could tell you what your risks will be before you join a crowd it has just scanned. "So, you know, do I really want to be entering into convention centers if they don't have a health monitoring system that can give us an indication or a reading of how sick the facility is?"

And that's what Chief Koskinas wanted to help accomplish. "Today, in reopening our communities and keeping our communities open, I need every possible tool," he told me. "I found it very appropriate for us to explore this."

And so the week of April 20, an FAA-licensed drone pilot began flying passes above citizens playing soccer in the park, standing in line outside the local Trader Joe's, and walking the streets of downtown Westport, not only to determine whether they were standing too close (the drone carried loudspeakers so it could tell people to leave a closed park or to maintain better social distance), but also to potentially identify symptoms of COVID-19.

Chief Koskinas explained that his department has adopted radical new technology in the past. Sandy Hook is a half-hour drive north of Westport, and the 2012 murder of twenty children there transformed his department. Before then, he told me, "we were preparing and training differently. Now it's common practice that a regular police officer carries a high-powered rifle every eight-hour shift." The drone pilot program seemed like something he *had* to do. "It might be something that there's no future to it. But not looking in that direction to protect our community? I think that would be irresponsible."

This particular project didn't last long. The day I reported on the drones for NBC News, the police department was so inundated with complaints that it canceled the partnership. People were deeply disturbed, a Draganfly representative told me later, to know that they were being recorded, and evaluated, by cameras flying above them.

But Chief Koskinas said he believes that his department may from now on have to monitor health as a matter of protecting and serving the residents of Westport, and so this test was an important part of that.

"So you can imagine a future in which a little league game, a US Open, a crowded train is scanned by your cameras?" I asked him. "And this technology would then tell you people are getting sick again?"

"Correct," he said. Then he paused, realizing that what he was proposing was a very strange new world. "There's the side of me, the human side, that looks at it and says 'this is very science fiction, or *Weird Science*' and it hurts. It really does."

We agreed that it felt like the plot of a movie we'd seen before. "But at the same time, it's reality," he said. "It's our new reality."

THAT NEW REALITY is taking shape. Decision-guidance systems are helping us to choose the music we play while preparing dinner, they're offering recommendations as to which people are the costliest to insure, they're telling us which route to drive between the airport and the hotel. How often do you stop cooking to change the recommended music or deviate from the recommended guidance on a map? These systems aren't just making recommendations; over time they've begun making decisions *for* us.

You can see the reasons why this is happening throughout our lives in some of the early adoption of these systems by our police forces and by the military. In those areas, surveillance technology is exploding. Why? First, the sheer efficiency of systems like AI-driven surveillance is deeply attractive to institutions like law enforcement, which are otherwise charged with labor-intensive, painstaking investigative work. The promise of AI is that its processing power can vaporize the up-all-night effort of conventional casework. And second, the companies that make these systems have struck on brilliant arguments that excuse us from any anxieties we may have about adopting them and their ethical implications. The invisible accelerating effects of these rationales are important to understand while we still have time to alter the progress of the technology.

Facial recognition, for instance, offers intoxicating convenience, while being extremely complicated to measure against our social values. I've interviewed the founders of many companies in the field, and time and again they've reassured me that facial recognition is being used in a very limited way. Modern facial-recognition systems have traditionally required a database of faces to compare yours against. And as a result, these founders have told me, their products only compare your face to a limited database of others, like a pool of employees, or mug shots from a single jurisdiction. "You can't just walk into a room and someone knows who you are," the vice president of a commercial facial-recognition company told me at the end of 2019. "That's called 'open set' facial recognition, and it's nearly impossible. No company in the US that we know of is doing it."

But it turns out that roughly three billion of us are already in a vast database of faces, one so large it might as well be open set, and any law enforcement agency interested in signing up has access to it. The database was built by Clearview AI, a small company first uncovered by the *New York Times*' Kashmir Hill, who found that it had contracts with police departments across the country. Clearview AI scrapes any publicly available photograph on the web for faces, and, when shown an individual portrait, can compare it with all the ones it has seen before.

It's startling to see yourself inside the system. Clearview's CEO, Hoan Ton-That, sat down in front of a laptop with me in March of 2020 and fed a small jpeg of my face into the program. It immediately spat out a pages-long grid of other photographs of me: shots from Facebook and Instagram, pictures from my appearances on television, me and a group of academics from my year at a university fellowship. And to test whether this was simply a function of my being an on-camera journalist, I clicked on the face of a woman standing with me in the academic group. Other photographs of her from across the web came up immediately. Click on any one of them, and the web page on which it was found appeared—everything an investigator would need to find out her name and affiliation.

Ton-That told me that his system is only built as an investigative aid, that it does not and should not constitute evidence of a crime, or be the impetus for an arrest.

"It's only a lead for an investigation," Ton-That said. "You still, as an investigator, have to match the name of the person, make sure they're in the right location, and then go to the judge and say, 'Look, I have evidence now.'"

But there's nothing built into his system that prevents law enforcement from doing what it likes with the information it gets. At the height of the drug wars of the 1970s and 1980s, the Drug Enforcement Administration repeatedly faced a dilemma: how to use information from confidential sources to prosecute drug crimes without having to compromise the source by revealing their names in court. And so began a legal tactic called parallel construction, in which, for instance, DEA investigators might learn a drug trafficker's identity from a confidential source, and then direct local police to follow that person's car until it rolled through a stop sign or blew past the speed limit. The resulting traffic stop results in a search, the search finds the drugs, the case is made. It's an easy way of doing exactly what Ton-That says one should never do. And the easiness of it is, of course, the problem.

That practice has since expanded to become a means of using ethically dubious surveillance technology as the foundation of an arrest. According to a senior DEA official quoted anonymously in 2013 reporting from the Reuters journalists John Shiffman and Kristina Cooke, "parallel construction is a law enforcement technique we use every day." The official called it "decades old, a bedrock concept." And that bedrock concept means investigators aren't just failing to resist the allure of something like facial recognition—they've built an entire ethical and procedural system for embracing it.

I asked Ton-That whether his system, and his company's willingness to give an account to any law-enforcement agency that asks for it, was simply too convenient for a police officer to resist misusing it in this way. He told me that administrators at a given department could limit their officers' use of the system, but that fundamentally his job is to build the tool, not the ethics to go with it.

"We expect them to use it responsibly," he told me. "And we want it to be used as much as possible within the policing world, because every crime that's solved, every murderer that's locked up is better for the whole society."

But what happens when The Loop spins the police round in a cycle of technology and shrinking choice? What happens to their choices when they have at their disposal the option to skip huge amounts of investigative legwork and just let the software do the identification for them? What happens when budgets and schedules for policing are built on the assumption that a software subscription can replace the need to pay overtime for detectives? These are questions we have to answer now. Because for one thing, this technology isn't just being brought to bear on the most heinous offenses. Three months after I met Ton-That, I interviewed Robert Williams, a Detroit-area father of two with no criminal history who was suddenly arrested on his front lawn, in front of his young children, and charged with robbing a watch store downtown. At the station he was shown a blurry photograph from the security camera on a printout from a facial-recognition system, and he snatched it up and held it next to his own face. "I hope you don't think all Black people look alike!" he said he told the officers. It was clear he wasn't their man, and he was eventually released. But his life was changed forever: even if he goes through a lengthy and difficult process of having his record expunged, his arrest could still come up in any future background search or job-application paperwork.

It turns out anyone who has ever posted their face on social media has essentially fed it into a public database for the police to use. The same month I spoke with Williams, New York activist Derrick Ingram joined a Manhattan Black Lives Matter protest on the occasion of Donald Trump's birthday. Ingram, who goes by D-Wreck, had regularly protested in high school, then fell out of the habit in his early twenties. Now he was back, with powerful feelings and a bullhorn. "That passion and anger came out on the megaphone that day," he told me from the courtyard of his apartment building. Photographs show Ingram on the far side of a police cordon, with the megaphone aimed at a female police officer standing just a foot or so away. Ingram says he doesn't remember the specifics of his encounter with the police—his case is still pending, and presumably he didn't want to risk compromising his defense strategy by sharing details with me. But the NYPD tweeted a photograph from the day, along with a picture of Ingram from Instagram, announcing that he was wanted for "assault on a police officer in the Midtown North precinct." They said he'd damaged the hearing of the officer by shouting at her through the megaphone at point-blank range. And weeks later, while Ingram was home in his Manhattan apartment, police arrived at his door.

He refused to let them in without a warrant, and a daylong standoff took place. "It escalated almost every thirty minutes," Ingram said. The police quickly displayed a knowledge of Ingram's background they could only have known by finding him on social media, the sort of thing Clearview makes possible. "They brought up my grandmother during it, saying, like, 'what would Delores think?'" Ingram's posts often referenced his protest group, Warriors in the Garden. "They were quoting me like, 'won't you be the warrior that you say you are?'"

In the end, the NYPD brought an estimated fifty officers to Ingram's home, plus drones and dogs. At one point, a camera crew caught an image of supervising officers carrying a flyer with Ingram's photograph from Instagram, underneath the words "facial identification section" and "lead report." Ingram did not open his door that day, and he streamed the standoff live. The next day he turned himself in.

An NYPD spokesman confirmed in a statement that facial recognition was used in Ingram's case, but that it's a "limited investigation tool, comparing a still image from a surveillance video to a pool of lawfully possessed arrest photos." Ingram told me he has never been arrested in New York, although he has been arrested in another state. The NYPD did not explain how it came by Ingram's photos on social media, although its public policy on facial recognition stipulates that "in rare cases . . . the Chief of Detectives or Deputy Commissioner of Intelligence and Counterterrorism may specifically authorize the comparison of an unidentified suspect's image against images other than mug shots, if there is a legitimate need to do so."

Six months later, after President Trump told a roaring crowd, "You have to show strength, you have to be strong," and that "we're going to have to fight much harder," hundreds in attendance forced their way inside the Capitol building—wounding and killing Capitol police officers in the process—and onto the floor of the House and Senate. That day, as thousands of photos and videos surfaced of people inside the building, I reached out to Ton-That to ask whether the FBI and the Department of Homeland Security were clients. He confirmed that they were. The next morning, I spoke to Sergeant Jason Webb of the Oxford, Alabama, Police Department, who described himself as "a power user of Clearview," and who, as part of a fusion center that brings federal, state, and local law enforcement together, said he was actively feeding the Capitol rioters' identities to his counterparts at the FBI, thanks to facial recognition. At that point, he told me, he'd already sent in between five and ten identities. "With the amount of videos that are out there and just the number of people that have their face uncovered," he said, "I feel positive that I can develop some more suspects."

I asked Ton-That in our first conversation together whether he thought people understood that his technology meant every photo they posted online was now available to law enforcement. He said he thought they did.

"They understand that when you're online, on social media, that people are watching," he said. "And you act very differently—than when you're around your friends or with your grandma."

I cut in. "I think if it were outwardly known to people that the photo they were taking of themselves at a party was going to wind up in a database used by law enforcement, they'd behave very differently than they do currently," I told him.

"Interesting," Ton-That said. "I think people would behave the same way, because they understand things are public."

But as the events of the summer, and the ransacking of the Capitol building went on to show, people *don't* know what's public. Most people truly have no idea that their behavior, their social network, their faces are all patterns ready to be recognized. They certainly don't seem to know that police departments are increasingly running suspects through facial-recognition systems. It may be that the Capitol rioters *should* be arrested using facial recognition. Maybe the same can be said of Derrick Ingram, who, according to the NYPD, permanently damaged an officer's hearing. But what about Robert Williams, who had done nothing wrong? We've allowed an industry of AI-driven systems to scan everything about us. Remember that early facial-recognition entrepreneur who told me that "open set" facial recognition was a long time out, ethically indefensible, no one's doing it? That was 2019. Just a year later, it turned out open set facial recognition is not only an existing product, the marketing of it is well-established, and it's being used all over the nation. The technology isn't just in beta. It's on the streets now.

Derrick Ingram says he first acted on the desire to make his political feelings public when he raised a Black power fist after winning a wrestling match as a fifteen-year-old. And his activism has been buoyed by the confidence that being out in public, expressing oneself, isn't a crime. But now he says he can see that police are suddenly empowered by technology like facial recognition, and aren't constrained in how they use it. "I felt like using high tech resources like that would be limited to people committing egregious crimes," he says. If he were fifteen years old today, he says, he'd be either less likely to post on social media, or less likely to be out in a public protest, or both. The technology has changed how police behave, and seems likely to change how we behave under surveillance. This is the beginning of how The Loop will reprogram us all, and we haven't even gotten to the part where it reads our everyday behavior.

Why do systems like facial recognition proliferate so quickly? Why doesn't public outrage or government regulation rein it in? Well, for one thing, it's moving too fast, and in darkness. Meredith Whittaker spent more than ten years at Google, where she worked on AI as founder of the company's open research group. Then she learned about something called Project Maven, in which Google was working for the Department of Defense. "This was a contract at the company that was applying Google's best-of-breed artificial intelligence in service of a DoD drone program—so using artificial intelligence to detect targets and surveil targets," she told me.

That project, along with revelations about sexism and racism inside Google, led Whittaker to help organize walkouts at the company. "A number of people in leadership positions were making choices about where certain technology was going to be used, how it was going to be used, and who would be harmed and who would benefit that many of us believed were unethical and frankly, unaccountable," she said. Whittaker resigned from the company in 2019, the same year Google announced it wouldn't be renewing Project Maven, and she went on to cofound and codirect the AI Now Institute at NYU.

She told me that there are several layers that shield systems like Clearview from scrutiny and outrage. First, the managerial structure by which technology is developed and deployed keeps people in the dark as to its ultimate uses. "A lot of times the people who are building an AI system, building an AI model, you know, creating the infrastructure on which these systems run, actually don't know how it's going to be deployed," she said. "They can make something that they may think is going to, say, help with disaster relief and find out later that it's actually deployed for surveillance or for other purposes." I'm often told by the public-relations arms of major tech companies that the technology that scans our behavior online is used to identify kidnapped children or sex-trafficking victims. But Whittaker quickly tosses that argument aside. Building a system that finds patterns in Venmo data to identify possible sex traffickers is, she points out, functionally identical to building a system that uses the same data to identify undocumented immigrants for deportation. The engineering is the same. The strategic *intent* is invisible to the people doing that work unless they ask—and they're rarely, if ever, allowed to ask.

Whittaker pointed to companies like Clearview and Anduril and Palantir, which make AI-powered surveillance and security technology for law enforcement and the military. There is no daylight, she argued, between the proprietary secrets of the companies as they develop the technology and the hidden uses to which they're put by law enforcement and the military. If the people *building* the systems don't know what they'll eventually be used for, you can imagine just how opaque it all is to those of us outside of the companies. There simply isn't any sort of public review. "Most of what we know about what they're doing and how it's being used have come through organizers and scholars filing Freedom of Information requests," Whittaker told me. "This is not something that is put to a democratic process to deliberate and debate. This is something that happens behind closed doors, and, frankly, is extraordinarily lucrative."

Spend even a day watching these technologies in action, and it's easy to see why military and law-enforcement agencies so readily buy them. In late summer of 2019, from a shaded observation platform overlooking a range of scrubby hills in Southern California, I watched a drone the shape of a miniature helicopter rise into the air. As it hovered at just a couple hundred feet above me, I couldn't hear the craft's rotors at all, and, as was pointed out to me, in darkness

I'd have no idea it was there. Anduril cofounder Brian Schimpf and I watched as the system showed us a close-up of a distant figure, captured by cameras atop several towers in front of us. The man was walking through the grass perhaps a mile away. The drone and the tower cameras locked onto the signature of his moving legs and swinging arms, and, swooping silently forward, began to follow him as he paced back and forth.

Anduril, founded in 2017, calls itself an "AI product company" devoted to making hardware and software specifically for national security. Its marquee product, Lattice, is a combination of drones and sensors that, Anduril claims, makes monitoring the edge of a military base, or the length of a border, an automatic process. "Instead of having someone post watch all night, staring at screens, you could automate this and alert them when there's activity that they should pay attention to," Schimpf said.

As usual with AI, it's about offloading the laborious System 1 work: "We can automate a lot of the more repetitive mechanical aspects and present the human with just the most critical information at the end of the day, so they can make an informed decision about what's happening."

According to a 2018 report from *Wired*'s Steven Levy, Customs and Border Patrol tested Lattice for ten weeks, and detected fifty-five people attempting to cross the US-Mexico border in the process. The company has been widely criticized by academics and watchdogs, including Whittaker, for selling technology that made policies like the Trump administration's family-separation policy easier to implement. (It's also been widely reported that Anduril went on to work on Project Maven after the internal protests at Google over the work.) But Schimpf told me there are some things the company won't do, that Anduril hasn't built facial recognition into the Lattice system, for instance, because the stakes for misidentifying someone are too high. Schimpf said he worried about what he called a "naïve deployment" of facial recognition, in which a company like his might compare a face to hundreds of millions of people. "Even if it's got a one percent rate of getting things wrong—"

"That's a lot of people," I offered.

"That's a lot of people," he agreed.

Inside the company's Irvine offices, in which engineers used hoverboards, scooters, and bikes to traverse the vast white working spaces, I experienced firsthand another way the component parts of the loop are being pulled together: the nationalist argument for getting on board with the deployment of pattern-recognition systems. Anduril cofounder Palmer Luckey told me that his company was necessary in part because China's technological sophistication and low cost of labor, combined with its lack of ethical restraint, meant the United States was at an inherent disadvantage.

"China is actually doing a better job than anyone in the United States for a few reasons," he said, not the least of which is that "China doesn't have any qualms about being a totalitarian police state." Luckey told me he doesn't *want* the United States to emulate that model, and that he founded Anduril in part to give the United States a chance to counter it, but that at the moment, China's lack of ethical concern about AI and surveillance has given it a head start in not only building surveillance technology, but exporting it.

"One of the things that is so brilliant about China's AI strategy is that they're not just building it for themselves," he continued. "In the same way that the soviets gave away boxes of AK-47s to other countries to get in bed with them, China is giving countries in Africa and Asia access to artificial intelligence technology that allows them to build totalitarian police states." In 2014, China's "smart cities" initiative began blanketing more than five hundred domestic cities in everything from traffic-monitoring sensors to cameras that feed images through police-monitored facial-recognition systems. And in the last few years, it has begun to export that technology to the capital cities of countries like Ecuador, Kazakhstan, and Nigeria.

"I think the United States needs to not just fight against that but also provide an alternate model," Luckey told me.

So what is the alternative? I asked him. What would a surveillance technology, infused with American civil liberties, look like?

"In that particular situation I don't know," Luckey admitted. "I'm more of a technology guy than a policy guy." (In my career I have heard a variation on that phrase a lot.) But eventually he settled on a notion. "I think the way is not through technological control. It's not to *not* have the technical capability. It's to have the political wherewithal to say, 'Yes, we know the tech exists to do this. But *you can't*. We're not going to accept that as a society.'" He suggested that countries like Mexico might accept a US-built system that offers extraordinary convenience but prohibits certain privacy infringements. "Facial recognition is becoming trivially easy. We can't make it not happen by not working on it. We have to say, 'The right way to control this is to set standards around it.'"

When I asked Whittaker about this notion, she pointed out that the sort of standards Luckey suggests have lagged far behind surveillance technology's tendency to metastasize beyond the first agencies and cases for which it was built. "You're not talking about a thin line in the sand where this technology stays. You're talking about this sort of bleeding into everyday life for most people," she said. "Police departments and other law enforcement agencies are now sharing these technologies, sharing databases, or sharing access to surveillance footage in ways that make it hard to draw a line between, say, the authority of immigration enforcement and the authority of your local police department."

At this point, the police and the military already have powerful surveillance technology at their disposal and are using it without almost any public oversight. The rare cases in which the people being monitored find out about it, and have the political clout to speak out about it, as the city of Westport did against Draganfly's pandemic-detecting drones, suggest there's a deep well of revulsion there. Anduril's drones, meanwhile, can silently and endlessly monitor anyone; they're just not deployed on suburban Americans. "Not yet, anyway," I said to Luckey. He disagreed with the insinuation. Luckey told me that working for government agencies puts natural guardrails around what he builds, because the military has higher ethical standards than the private sector does. "I would feel differently if we were selling drones to consumers," he said.

"If we were just building our stuff and saying, 'Hey, Joe Smith can buy our towers and buy our drones, and he can set up a surveillance system around his ranch,' that would be very different."

I've replayed my conversation with Luckey many times in my head, and I honestly don't know how I feel about it. He may very well be right that the military holds itself to higher standards, imposes true ethical boundaries, as compared with the companies I've spent so many years interviewing. Certainly he's right that the improvisational process by which I've watched companies apply technology to analyzing and shaping human behavior, rationalizing their ad hoc policies using either shareholder value or ever-shifting corporate principles, is not adequate to the outrageous influence they've come to wield over our lives. But the military and law enforcement are specifically charged with matters of life and death, liberty and confinement. When they bring a product like AI to bear on identifying the cadence of a human pair of legs, or predicting the whereabouts of an undocumented immigrant based on their financial history, the point isn't to sell that person insurance. It's to kill or capture them. Ultimately, it's difficult to trust either private or military use of pattern-recognition technology when you and I don't get to see how it's used, much less consent to its use, even after it's been used on us. And yet here it comes, into our lives.

THE DEBATE ABOUT military and police use of AI and surveillance technology is the most visible edge of The Loop. AI-powered robots that conquer humanity by force are a standard trope of dystopian science fiction, and when I describe The Loop, people often assume I'm talking about *The Terminator.* But I don't worry about robot overlords enslaving us. I worry about us using robots to amplify the wrong part of our instincts, such that we wind up enslaved to *those.* So I want to turn now to another area where The Loop is taking form, but in a part of our lives where it's much harder to put our finger on what exactly is at stake. And that's childhood.

It takes a long time to study human behavior. Or more importantly, it takes a lot longer to study human behavior than it does to change human behavior. Even the most reputable academics need a couple of years of grant writing to score the necessary federal funding to study something, and then a few more years to conduct the study, and then a few more years for the study to be accepted and published by a peer-reviewed journal, and then it takes all that time again for someone else to independently replicate those findings until something like scientific consensus is eventually achieved. A whole generation of kids can go by before a population-wide effect is even close to established. Consider, by comparison, how fast screens have hit society.

Just fifteen years after the iPhone and ten years after tablets were introduced, the average American child now experiences her first screen at about four months of age. The estimate is that somewhere between 40 percent and 50 percent of American children under the age of five now own their own tablet or phone, which may account for the estimate that those children are spending an average of three hours a day on the devices. The *behavior* is here. How do we study it?

A small group of pediatric researchers is moving as fast as they can to get out in front of these new screen habits, trying to sort out how they may be changing the child, and the child's family, and what, if anything, we can all do to reconnect with one another.

I walked into a downtown Seattle office tower one January before the pandemic, and like nearly everyone else in the city that dark morning, slugged a coffee from the hissing espresso machine downstairs before heading to my first meeting of the day. Upstairs I walked through several corridors and shook hands with Dr. Dimitri Christakis, one of the authors of the current national pediatric guidelines about screen use.

Christakis is a rapid talker, and quick to laugh, but his obvious fascination with what he's been observing over the last decade is weighed down by what seems to be a growing realization that it's mostly bad news. We sat in a dark observation room and watched a video of a young child, maybe a year and a half old, enrapt in a tablet in the next room. Christakis rapidly explained the dilemma to me. It begins with just how much faster technology moves than good science can.

"When I'm wearing my scientist hat, I have the equipoise that all scientists should have," he says. "I can become very mealy-mouthed and say, 'We need more research, and we can't make any conclusive decisions.'"

"But when I'm wearing my clinician hat," he continues, "and I'm in a room with a family and they're asking me, 'How much time should my toddler spend with an iPad?' Or, 'I'm really worried that my child is showing signs of addiction.' They don't want to hear from me 'I've got good news! I just started a research study, and two years from now I can tell you what to do with your two-year-old.'"

Christakis has studied all ages of children, from babies to teenagers, and how technology, from television to phones, affects their health. But in recent years, he has become focused on the earliest possible point at which a child might develop a compulsion around technology. And he's trying to create a screening method for spotting that compulsion.

What's the difference between a child who just loves to play with an iPad, and one who has a compulsive relationship to it? Most researchers hate this kind of question, and they fill their answers with qualifications and verbal footnotes to avoid oversimplifying anything. But Christakis doesn't hesitate.

"The typical eighteen- to twenty-four-month-old will spend an average twenty to thirty minutes with their favorite toy, whether it's blocks or books or dolls or trucks. That's it," he says. "We know now that children will spend much, much more time with iPads, with touchscreen media. And that alone tells us that there's something fundamentally different about this experience. Many parents think of it as a digital toy. It's very, very different than a toy. There's something about it that's sufficiently compelling that children play with it in a way that they don't play with other toys." I think of Wilkinson pouring $50,000 into a dead-end casino simulator on her phone. I think of Mak Kapetanovic as a lonely teenager spending hours following the algorithmic flow of YouTube videos on "race realism." I think of the roomful of young entrepreneurs listening to the two addiction experts describe the human brain's ability to rationalize any addiction.

Remember the "machine zone" Natasha Dow Schüll discovered so many slot-machine addicts were after? Christakis says that not only does the unusual length of time a child becomes engrossed suggest compulsion, but the emotional character of that extended time on an iPad suggests that children are leaving the realm of playful joy behind. Their countenance becomes grim, they don't smile, their manner doesn't suggest a human experiencing a full range of emotions.

At his lab, we watch footage together of a young girl, perhaps two and a half, with her mother, and Christakis watches for one of several possible measures his project may suggest for something like a national screening program. The researcher on the other end of the table gives the little girl a succession of activities: first an electronic toy shaped like a guitar, then a tablet loaded with a guitar simulator. In both cases, she occasionally interrupts the girl with a pointed finger and an urgent "Look!" The little girl, an upbeat, outgoing kid in a black sweater and white tutu, each time abandons what she's doing and cranes around eagerly to see what the researcher is pointing out. By the third time, she also responds with a pointed finger, mirroring the gesture.

Christakis nods in recognition. "This is what's called dyadic attention," he says. It's an evolutionary gift all healthy children have. "They're doing that, actually, in an effort to learn," he says. "They're assuming that you're pointing out something to them that's important."

And the little girl pointing out things to the researcher is the other half of the dyad. "As children get older, when they get to be about a year of age, they will orient parents. They'll point to something, and a parent will instinctively look at it, and then look back. That exchange implicitly asks the question, 'What is that? Tell me something about it.' It happens twenty, thirty, a hundred times a day in normal parent/child interactions."

Then we watch the little girl turn her attention to her favorite iPad game, an app in which the *Sesame Street* character Elmo urges you to draw the letters of the alphabet. "Elmo!" she whispers to the researcher with a faint smile.

This time, when she's ensconced in the game, scribbling at the letters, and then gravely watching the short videos that reward her, the researcher again points to the back of the room. "Look!" she says firmly. The little girl doesn't notice. Her eyes are fixed on Elmo.

Christakis shakes his head and smiles. This is not what children do under normal circumstances. "It's unusual," he says. "They're overriding what's a really strong instinct to look. And it's a sign that whatever they're doing now is much stronger than their natural instinct."

There are some human instincts we want to move beyond, certainly, but this isn't one of them. Christakis points out that dyadic attention, passed from generation to generation, is considered a crucial part of a child's development. "Imagine this," he says. "The parent could be pointing out something that's dangerous." A snake. Fire. The engrossing nature of the iPad has robbed this little girl of the helpful power of one of the most fundamental human instincts. Instead, Elmo has her complete attention.

Is it somehow better that this is an educational app, with a trusted and beloved children's television character inspiring the girl to try to write her ABCs? Christakis doesn't believe apps have been studied well enough to know that it makes a difference.

"It's not entirely inconceivable that an app could be educational *and* addictive, right?" he asks. Children's television has walked this line for two generations, trying to make something like *Sesame Street* both engrossing and redeeming. And certainly the way my children read certain books could be described as compulsive. They wander the house with them, bumping into furniture, emerging from the restroom without any pants, head down in the page. But Christakis says that the compulsion-forming potential of apps is taking a generation of parents by surprise, and that he worries they're blind to the compulsive effects they may be having. "I don't think they think of it as an addictive substance," he says. "I think they think of it as either a neutral product, or as an educational product."

And unlike books and *Sesame Street*, he doesn't believe there's any real evidence that apps are educational. "The vast majority of apps, which is to say over 95 percent of them that are designed for toddlers and young children, are advertised as being educational. They claim either explicitly or implicitly that they will teach letters, numbers, social skills, music. And the overwhelming majority, over 90 percent of them, have absolutely no evidence base to support those claims. In fact, in many cases, there isn't even a theoretical basis to believe that they could deliver that," he says. "So from my perspective if an app claims that it is educational it should be obligated to prove that it is. And it could do that. And that is something that could be regulated. But it currently is not."

FROM MILITARY AND policing to childhood development, it's increasingly clear that technology shapes and even creates our unconscious behavior, that we don't naturally possess the mental or social tools to resist it, that we need to construct systems of self-control to counter the systems we've built to hijack it. What would such resistance systems look like?

About an hour north of Lansing, Michigan, just inside the entrance to Central Michigan University is the football stadium, where an enormous red sign reads FIRE UP, CHIPS! with a picture of a charging player. It's an unfortunate reference to one of the area's ugliest legacies: the school's mascot is the Chippewa, the English name for the Ojibwa tribe forced out once settlers discovered the excellent white pine here. The school was founded in the 1890s to train teachers. But just across the street from the stadium's reminder of a dark history sits an extension of one of the school's brightest accomplishments: a cutting-edge research clinic for parents and educators.

I arrived early on a weekday to visit a program that specializes in coaching parents using a system called Parent-Child Interaction Therapy, or PCIT. The logistics involve a parent and child on one side of two-way glass, and an expert on the other. The parent wears an earpiece and hears the voice of the expert coaching her or him in real time.

In its forty years, PCIT has become an extraordinary way of improving family relationships, especially when something big is in the way. Incarcerated parents, who get only periodic visits with their children, use it to rapidly acquire family and parenting skills after their release. Parents who have spent a decade addicted to a substance like methamphetamines, and are now sober, use PCIT to catch up on those lost years. Parents of children on the neurodiversity spectrum can test-drive the latest methods under the immediate, invisible supervision of an expert.

But psychologists Larissa Niec and Sarah Domoff have adapted PCIT for a new era, and I was here to witness it. They've helped families with difficulties around mealtime, physical outbursts, and other behavior. And now they've adapted the program for families in which a child has an addictive relationship to technology.

That morning, I met a couple, Andrew and Heather, and their six-year-old, Miles. The three of them had completed thirteen weeks of the program and were here to model it for us. Miles is a big, strong kid, with a cowlick at the peak of his sandy blond hair that bobs behind him when he runs.

As Miles ate pizza in the next room, I asked Andrew and Heather what brought them here originally. Andrew sighed.

"It was the transitions, I think, that were the hardest, where he would want to play a game or want to watch a movie, and he would do that, but of course the time would come for it to be over. And when that time came, it was very rough."

How rough?

"He was throwing rocks at the windows in the backyard," Andrew said.

"That was the moment," Heather told me, "when we said 'we need help.'"

Domoff had done several studies of the effects of screens on children, from toddlers through high school, when she came together with Niec to adapt PCIT for the technology era, naming something new in the process. The program is called PATCH—Parents

Active in Their Child's Health. "It's really focused on taking what we know works for managing child behavior in other contexts and applying it to these new contexts that parents are facing today," Domoff says.

Niec and Domoff began circulating simple flyers at pediatricians' offices in the region:

> Is your child's behavior hard to manage?
> Are meal times a struggle?
> Do you have trouble dealing with your child's
> use of tablets, video games, TV?
> The PATCH program may be for you!

They decided to work with children between the ages of two and a half and seven, and offered the participants a small stipend. They say they'd hoped they'd get a few dozen families. Today, hundreds have applied.

Miles sat down in the playroom with his mother, and they leaned together at the table over an intricate model train set. Domoff, Niec, and I watched from the observation room, and Niec picked up a microphone and began coaching Heather in the gentle voice of a children's book narrator.

Heather had learned from her training here to praise Miles in a specific, directed, and extremely pleasant way. When Miles picked up the locomotive and handed it to his mother, his mother kept it simple: "That's the train engine! How cool!" We could hear her and Miles on small speakers in our dark room. Niec immediately piped up. "That's great," she says. "Really good labeling, not asking questions, just showing him you've heard him." And when Heather announced that soon she and Miles were going to be able to watch a video together on her phone, and that they'd watch just a little of it, and then go back to playing together, Niec nodded happily. "Great time management skills there, great job," she told Heather.

And then it was time to use a screen together. This is the moment everyone in the clinic had been anticipating. I thought of my own kids, and how committing to a video together means rolling the dice on a whole afternoon's worth of potential meltdowns. Miles and his mom sat together in silence and watched. Miles was entirely, unnaturally still. His cowlick never moved. Then Heather pointed out a detail here, or a character there, and Miles began to narrate along with her. Over the earpiece, Niec suggested to Heather that she give Miles a heads-up about winding the video down. I stiffened.

"We're going to give it another minute, and then we need to move on to something else," Heather said, leaning in close. "But you and I are going to get to play a little more, too, okay?"

Miles nodded.

"Nice job with that preparation," Niec said. "And see, he accepted that preparation from you!"

And then Heather asked her son to hand back her phone. There was a moment of hesitation, and then this kid, who used to scream and break things at moments like these, gave it to her.

"You did it!" Heather said. "Even though you didn't want to, but I really appreciate that you did it."

Miles leaned over and silently kissed his mom on her shoulder.

I looked at Domoff with my eyes wide. She was already nodding at me.

"They don't want the screen, they want *you*," she whispered.

Heather told me later that this is what they have taken away from the experience: "Making sure that instead of being like, 'We're going to use this screen time so that you're good, and I'll give you time later—'"

"It's reprioritizing," Andrew said.

Heather nodded. "By giving our son what he needed first, he's better able to manage the transitions." She thought on it for a moment. "His needs have been met, and so it does honestly make other times easier."

Watching this process, I had the clear feeling of discovering a sort of hidden training that should be available to everyone. Any parent can tell you that being sent home from the hospital that first day involves the feeling that someone must have misunderstood, we are in no position to raise a baby by ourselves, are you kidding me? We have no training!

Here, for me, was the training. Who wouldn't benefit from having a coach in their ear during a dozen practice rounds of dealing with a difficult thing my kids do? And meanwhile there is so much inadvertent, counterproductive training we go through as parents and children in a world where screens have been normalized by their constant presence. According to a June 2021 report from Deloitte, the average US household now has a total of *twenty-five* connected devices. Several parents I spoke with told me they get as much guidance on the proper role of screens from the ads they see by the companies who make them as they do from anyone with actual expertise into how the technology might affect their families and their relationships. Only parents who happen to notice the right flyer in central Michigan, or Seattle, might be getting any professional guidance about the effects all those screens are having on children (or on parents, for that matter).

And at the same time, the ancient systems we have for passing information between parents and children powerfully shape our kids. If parents are infected with new behavior, it spreads quickly to the children. What are children learning about the expected role of technology in their lives after a pandemic that had them watching their parents more or less constantly connected to a device of some sort?

Developmental researchers have long understood that we use our social sensitivities to offload decision-making from our brains to the brains of others, and that the offloading from parents to children is especially powerful. In a classic 1960 study, psychologists Eleanor Gibson and Richard Walk built a visual cliff—in essence, a high box, more than four feet tall. Where it dropped off on one side, Gibson and Walk laid a thick sheet of glass, capable of supporting a good deal of weight. They then brought in thirty-six infants between six and fourteen months old, and had the babies' mothers call them from the box out across the glass.

The results were an embarrassment for humans. Only nine of the children did the sensible thing and refused to venture out over the glass. Fully twenty-seven other children willingly crawled across what looked like a sheer cliff. The researchers brought in dozens of other baby animals to compare their performance. In the case of some species, like goats, not a single one would willingly risk the glass.

And then, in 1985, a team led by James Sorce tried a subtler variation of the experiment, drawing on the idea that what the parents modeled for the children might make the difference. They built a visual cliff that presented greater uncertainty, by being a little less daunting—only roughly thirty inches high. The researchers used only one-year-old subjects. And this time, they asked the mothers to act out a variety of emotions in response to the cliff, signaling approval with big smiles, or changing their faces at a certain point to display intense fear. When the mothers looked happy, fourteen of nineteen babies willingly crawled forward over the edge. But when the mothers looked afraid, none of the babies risked the glass. Most of them backed away from the edge.

Take the powerful, almost narcotic effects that researchers like Christakis, Domoff, and Niec are working to quantify and struggle against. Couple those with the long-term effects that watching a parent absorbed in a device may have on young children trying to sort out their own relationship to a screen. Consider how powerful childhood memories are in determining what that child will consider appropriate and inappropriate behavior as a parent later in life. And think about how long it will take before the research will be there to establish any causation, much less any parenting strategies or federal safety regulations, even as generations of children are changed. As adults, we barely register the creeping effects of technology like surveillance and facial recognition, and how they might change our expectations about privacy and public life. Meanwhile, children are growing up in that world, exposed to all these effects from the day they're born, without any guidance from their parents as to how it might be shaping us all. This is The Loop gathering momentum, whirling powerfully just outside our immediate ability to sense it. And it's beginning to gather up not just our privacy, not just our family time, but everything else, too.

Chapter 9

THE LOOP

UNTIL THIS MOMENT in history, the discoveries of behavioral, political, and social researchers that I've described in the first chapters of this book were mostly a matter of pure science, the foundation of pivotal, abstract academic theories, maybe even the basis for a handful of new social programs and evidence-based systems. Marketers have co-opted the broad strokes of those findings, political campaigns have thrown some of it against the wall and found that some of it sticks, but as we've looked into the statistical patterns of how humans make decisions, studying those patterns hadn't been used to *change* human behavior. It was mostly a matter of investigation, debate, refinement, not work that touched down in our lives in any tangible way. Now, with AI, all of that unconscious behavior—the tribal instincts, the compulsions, the inability to parse true risk, the sacrificing of long-term benefit for short-term diversion—is data, and the foundation for wildly profitable business. As a cloud of variables, it gives machine learning an endless opportunity to find patterns and predict what we're likely to like, click on, buy, join—to *do*—next. And as we've seen, those choices are not only unconscious, they camouflage themselves as rational, reasonable, rewarding courses of action. They feel like the right things to do, even when they are not. As a result, a system built to find the patterns in our instincts, to detect the movies that we enjoy, to connect us to people with whom the data suggests we have common cause, and to make all these suggestions in ways that flatter our tendency to think well of ourselves, is not only irresistible to us, it's also an extraordinary business opportunity. Meanwhile, an almost total regulatory blind spot for behavior-monitoring technology means the companies that build them aren't meaningfully answerable to any oversight body. We make choices that feed these systems data, the data is examined, the data shapes our choices, our choices feed in more data. This process—a powerful, circular current of unconscious human instinct, an inability to see when we're being swept around and around inside it, and the capitalist instinct to make money convincing one another to stop paddling altogether—forms The Loop. And here I'm going to show you how its third and most powerful ring is beginning to encircle us.

At age eighteen, Anna Todd married her high school sweetheart, Jordan, an army soldier, and left Dayton, Ohio, with him for his post at Fort Hood, Texas. Opportunities were few, and life began to look very predictable. She waited tables at a Waffle House. She worked a makeup counter. And once she had a son, Asher, the cost of childcare meant staying home with him made more sense than working any longer.

"He's a special-needs child, and we were just kind of hanging out during the day," Todd told me. "He has epilepsy, and he was having a lot of seizures then. So I would just read, give him medicine, read some more."

Todd was really into romance novels. But not the traditional paperback form. She read them in digital installments, as new chapters were released on a free app called Wattpad. But soon Todd began to notice a dissatisfying cycle there, too. "Every story was 'they fell in love, had one conflict, resolved it, happily ever after,'" she remembers. "So I thought 'oh, I'll just write something to entertain myself until someone else writes this book I'm looking for.' I didn't realize that I was going to be the one writing the whole thing."

Todd began writing her own stories on her phone standing in line at the store, sitting in a dentist's waiting room, feeding Asher with one hand as she typed with the other.

"I didn't have any reason to think that what I was going to write anyone would read, or that it would become more than one chapter," she said. "I only thought about that one chapter. I never thought about anything beyond that." But when she released her first installment on Wattpad, readers started commenting on it, and suddenly a community began to form around her work, with readers talking to one another about her story, comparing their impressions. For someone living a very lonely existence, and who was deeply connected to the world of romantic fiction, this was a powerful experience. So she kept going.

Her first fully formed novel was a dark take on young-adult romance, in this case between a college freshman and a rebellious fellow student (modeled on the lead singer of the band One Direction, on whom Todd had a crush), and she released it the way she'd seen others do it, a chapter at a time, on Wattpad. Her readership kept growing. Wattpad's interface changed, such that readers could now comment not just chapter by chapter, but line by line. At first, she responded eagerly to every comment. But soon so many people were cheering her on, asking questions, making suggestions, she says, "there was a day when I couldn't write back anymore."

Soon Todd began receiving emails through the app from people claiming to be literary agents. "But I ignored them," she said, worrying that they were grifters or stalkers, and feeling embarrassed at the prospect of being identified as an erotic author.

Then emails began to arrive from the employees of Wattpad itself. "They said, basically, 'your story is breaking our site, and we'd love to know more about you,'" she remembers. "Things like 'We want to know where you're from, how old are you, what do you want to do with your story?'"

They also told her that they had publishers lined up, ready to turn her work into a novel. Now she believed them. They flew her to Los Angeles, and then to New York, where she met with several houses. And by 2014, less than a year after she'd begun writing in earnest, she had a half-million-dollar multivolume publishing deal. She went on to write several sequels to her first book, *After*. When I met her, at Wattpad's office in Toronto in 2019, Netflix had just released a movie adaptation of the first book, and she was dressed in black for the premiere party.

Todd says other writers now regularly ask her how she broke through on Wattpad, what the secret is to being noticed in the crowd of aspiring authors gathered there. But it turns out Todd doesn't know what the secret is—"I wish I knew!"—and she's never had to so much as edit her work to please her publishing and film partners. "I just got to write whatever I wanted," she said. "The only things I did that I would tell people to do, which I didn't know I was doing right at the time, is engagement, and just kind of building a community."

There are other, hidden factors that contributed to her success. It turns out that Todd's work appealed not only to readers, but also to pattern-recognition algorithms. That's a big part of what Wattpad uses to spot potential hits. And perhaps more important, her eagerness to connect with her readers helped to form a community that fed additional data Wattpad's system needed to identify whether her work was going somewhere.

Wattpad's business model, and Todd's extraordinary success within it, are an example of a new form of algorithm-powered cultural curation. On the one hand, it's making stars out of people like Todd, who would never have gotten a shot at writing a book or optioning a film idea before now. On the other, it plays right into a series of statistical illusions that put the publishing and film industry onto The Loop, feeding familiar, successful stories back to us ad nauseam.

Spotting success and repeating the things that look like its ingredients is a trap we've fallen into before. Publishing and film studios already operate on this model, with humans, rather than algorithms, selecting likely hits and greenlighting sequels. But the tendency goes back a lot further than that. In a 1943 report for the Statistical Research Group,[1] the Hungarian mathematician Abraham Wald sought to answer an essential question for the US military: What part of combat aircraft needed to be best protected by armor? The damage from explosive shells, flak, and machine-gun fire was obvious in returning planes: the wings and tails of aircraft were riddled with holes. Logically, those parts of the planes needed better protection.

But in a series of memos for SRG, Wald revealed that it was the *downed* aircraft—the planes that hadn't returned, with damage one could *not* see—that had the data the War Department needed. He realized that the damage that could be seen in returning planes was the kind that still allowed a plane to return home. And by instead measuring what could *not* be seen, he showed that the damage most rarely seen in returning planes—7.9mm machine-gun fire across the cockpit, 20mm cannon damage to the engine—was most likely to bring planes down. His findings led to reinforced engine compartments (and presumably the survival of untold numbers of air crews) and transformed American aviation logistics during World War II, the Vietnam War, and the Korean War.

Wald is credited with helping to identify something that statisticians now work hard to fight off: *survivorship bias*, the tendency to optimize our behavior for the future based on what may in fact be the rare instances in which someone or something survives extremely long odds and only as a result comes to our attention. Forecasting probability based on limited data is something statisticians now know not to do, thanks to Wald—data scientists often post his report's famous diagram of a shot-up plane as an online meme—but Anna Todd is the spear-tip of an industry that forecasts future success by studying survivors.

When Todd and I met, roughly four million writers were posting more than twenty-four hours of reading material to WattPad every sixty seconds of every day, according to the company. That's a lot of artists looking for an audience, much less hoping to be discovered by someone who will pay for their work. And yet some had, in fact, been signed. As of 2019, the company told me, nearly one thousand Wattpad stories had become published books, TV shows, films, or some other form of digital project. That number is one that a magazine or book publisher would be proud of even two decades ago. But as a percentage of total writers, one's chances of breaking through with a book or script deal are tiny: just 0.00017 percent of more than 565 million uploads. Todd is the most successful of that tiny fraction of a percent.

Allen Lau, the CEO of Wattpad, told me that machine learning is the key to finding authors like Todd. "Machines are very, very good at analyzing massive, massive amounts of data. And this is what we have." And with the rise of natural-language processing, in which an algorithm can glean bits of meaning from what people write, Wattpad can scan not just the writing itself, but what's being written *about* the writing. "We are seeing hundreds of millions of comments on those stories every single month. So, a lot of the insights, a lot of the emotion, is actually contained in the comments," Lau said. Patterns in the comments, he explained, point the company at the stories that appear to trigger real loyalty in an audience. "This type of technology is very, very good at identifying amazing content [in a way] that wasn't humanly possible in the past," he said. His algorithms look at the path the readers took to that particular story, how long they read, how fast they return, how rapidly they're commenting, and the language of both the writing and the comments.

Aron Levitz, the general manager of Wattpad Studios, sells the platform's best performers to publishers, producers, and agents. "We're using some form of data to find the writer first," he told me. "Now, to be clear, though, audience data is only half of it. It's also being able to look at the pattern in the words." Wattpad has optimized the algorithm to find a certain emotional resonance and echoes of past hits. "What are the emotions it's bringing out? Oh, this is about wizards? Oh, this is about schools? Maybe it's a little bit like another story we've heard of in the past." He smiled.

I asked Levitz whether there's any danger that audience data, plus content analysis, will result in a system that simply serves up minor variations on the same content over and over again. I'm trying to ask him about the risk that rather than find *new* content, he'll simply create an endless cycle of the same old content. In other words: How will he avoid The Loop? He insists that it's not that the stories themselves are the same as past stories. It's the effect discernible in the audience. The AI detects patterns in how people *react* to the stories that are similar to past reactions to past stories.

AI isn't everything, he said. "It doesn't replace human judgment, but with the data backing this up, we will know more reasons why this [story] is taking off." The company essentially tells its partners in the publishing and film worlds that a piece of software, trained on the indicators of past hits, has endorsed a particular work—and earlier in its life than any prior system was able to do. "We are reducing the risk. We have much more confidence in this piece of software, and we have much more confidence in this piece of content before someone makes a $20 million decision to turn this into a movie."

On the face of it, Wattpad is, in fact, finding *new* content. The stories the algorithm chooses come from all over the world, often from cultures and languages that have been largely overlooked by Western audiences for years. It may very well be that the algorithm will diversify a system that has been based on the instincts of mostly white, male gatekeepers.

But again, we're here looking at these trends together not just to see whether a company can use AI to diversify our lives marginally in a few financial quarters. That can undoubtedly happen. What we're here to examine is what market forces and technological capabilities and hardwired human tendencies are going to come together in the next few years to shape how we all live in the next one hundred. Again, The Loop hasn't fully clicked together yet. But the pieces are sliding toward one another, and in Wattpad I see an enormous chunk of it built already.

If, as Wattpad's executives told me, the audience's reaction to a piece of content is something they can measure, my worry is that the arc of approval for that piece of content will look the same to AI, as a pattern of data, no matter the writer or her background. This then runs the risk that we'll create a pattern-recognition loop that traps us in a world where the only films and books that get greenlit are the ones that consistently provoke the same reactions in an audience. How will we take the occasional flyer on the unexpected cultural hit—an impressionist painter, a Julia Child, an *Arrested Development*—that people like for reasons they can't describe, if it requires an algorithm's signoff to even bring in a concept for human consideration?

And what will an algorithm that looks for the same broad emotional and energetic qualities in human reactions wind up choosing over and over again? Already, we're beginning to see what algorithmic curation produces on, say, social media. As the journalist Kat Tenbarge wrote in an essay about the blandness of the influencers that rise to the top of TikTok, "Teens are plucked from obscurity by an algorithm to become overnight influencers. It's a cycle that has already bred endless controversy, pervasive relevancy, and . . . invasive mediocrity."[2] While TikTok is a showcase of exceptional creativity from all over the world, its most famous performers often don't stand out in any particular way—they're beautiful, but they don't shine as dancers or singers, and they're usually mimicking moves invented by other creators, and moving their lips to songs written and performed by other singers. And yet these are the performers seen by millions and paid millions. They're the shining stars of the data set.

In 2021, Wattpad was acquired by Naver, the search-engine giant of Korea. Naver's business is like Google's—search, audience analysis, advertising—but rather than put the profits into things like automated thermostats or drone-delivery systems, as Google's parent company has, Naver has invested in a very different kind of product. They find and promote and export cultural content, from music to scripted dramas to graphic novels. Naver's CEO, Han Seongsook, told me from Korea that the company had already been using pattern-recognition systems for years in its Naver Webtoon business, the world's largest online comic and graphic-novel platform, and that Wattpad will be another verdant garden of patterns when it comes to what audiences like. And Naver Webtoon's founder, JunKoo Kim, told me that he was experimenting with AI-driven systems that could actually *draw* a graphic novel *for you*. "You don't need to be a great artist, you can just tell the software what you want," he said proudly. Already, artificial intelligence is being used to not just choose the creative work we'll enjoy, but to make it.

Here, of course, anyone of even the slightest philosophical bent will lean in and say, "Wait a minute, what do you mean by 'make'?" And of course, that's the right question. Because as we know, AI cannot execute spontaneous creativity—it can only draw patterns from what has come before. In 2019, the Harvard philosophy professor Sean Dorrance Kelly wrote an article for the *MIT Technology Review* in which he argued that AI cannot be an artist.

> We can't count the monkey at a typewriter who accidentally types out Othello as a great creative playwright. If there is greatness in the product, it is only an accident. We may be able to see a machine's product as great, but if we know that the output is merely the result of some arbitrary act or algorithmic formalism, we cannot accept it as the expression of a vision for human good.
>
> For this reason, it seems to me, nothing but another human being can properly be understood as a genuinely creative artist.[3]

In an abstract sense, that's correct. If we were to haul a piece of AI in front of a jury and cross-examine it about its abilities, we'd quickly see it's just a simulation. But it turns out that if you haul a piece of AI in front of a roomful of art buyers at an auction house, they'll buy what it makes.

In 2019, the German artist Mario Klingemann talked me through his newest work, *Memories of Passersby I*, in a London gallery. It's a pair of sixty-five-inch screens oriented vertically on a wall, connected by wiring to a chestnut TV-style console on the floor. Inside the console sits a computer running AI that generates an ongoing, never-ending, never-repeating medley of woozy, impressionistic oil-style portraits, one face on each screen.

Klingemann spent months feeding the AI his taste in painting—specifically, he fed it more than 100,000 seventeenth- to nineteenth-century European oil paintings, and then told it which ones he liked best. He used generative adversarial networks, or GANs, in the work, which means that one portion of the program (the generative portion) generates image after image, while another (the adversarial portion) lets through only the images that hew to the patterns it has detected in the paintings he liked. "You give it a training set," Klingemann told me, "and it will create new images that are statistically similar to what you fed it."

At first, Klingemann said, he had to review what the GANs let through, because they would begin to drift into this or that tendency he didn't like, and he'd have to tweak the output. But soon he could just let it go. "It's a bit like cooking," he explained. "You start doing it, and then you say 'too hot' or 'too sweet.' And at some point you get something that looks right." Soon, he was satisfied. Or at the very least, he thought he was. Even in traditional painting, he pointed out, "there's a certain experience that's out of your control. It's really like you know it when you see it." With AI, he said, "it moves so fast." But before he knew it, he had something he liked. "Sometimes you didn't know you wanted to go there, but then you go there and it's better." Better than what? Better than he planned on. Better, perhaps, than what he could have done himself.

His audience seemed to agree. *Memories of Passersby I* sold at a Sotheby's auction for £40,000. It might be that the appetite for these AI-derived pieces of art is simply an appetite for novelty. Maybe that's the ingredient that has made them a valuable commodity. I struggled with that notion for a while, feeling tempted to dismiss the whole thing. But then along came NFTs.

NFTs, or non-fungible tokens, are pieces of code slipped inside any digital file that identify it as "unique" in the open-receipt system known as blockchain. Once an NFT has been placed inside something like a movie or audio file, that particular file—just the one—is, in theory, the only one of its kind, and in the public ledger of the blockchain, anyone can look up that particular NFT to see whether it's registered to a particular person. This allows humans to own a digital file in a way that no one else—even if they have a file that's identical in all other ways—can claim.

Justin Blau, the DJ and producer known as 3LAU, told me from his home in Las Vegas that in ten years of touring, he made roughly $12.4 million. In the first four months of 2021, selling NFTs to fans who wanted to own a limited-edition version of his work, he made $12.9 million. But why? I asked him. What possible reason do people have for spending money on something they can ostensibly have for free?

"It makes them feel different than if they were just listening to it on their own," he told me. "It's a feeling. People buy experiences, not just products and services. People buy experiences because our entire lives are shaped by our memories. And those experiences give us happiness."

"Ownership of that art creates immense value in other people's lives," Blau said. "It's an emotion that I feel. And that's what people are buying here."

There's clearly a deep hunger for the added layer of emotional satisfaction that we seem to feel about *owning* a piece of art or music, even in an ephemeral, digital, ostensibly nonsensical way. I think back to Kahneman and Tversky and Slovic and Thaler, the work they did on our deep emotional allergy to uncertainty, on our unconscious need for a reliable outcome, on the imbuing of things that we own with some special, magical quality *just because we own them*. It's possible that all this boom-time digital art is simply the product of vast wealth inequality (and that's definitely a factor) or of novelty (that's undoubtedly one as well). And it may be that NFTs somehow offset the algorithmic dilution of art in all its forms, by putting pressure on artists to differentiate themselves, and each piece, so they can charge for their work in a landscape that has otherwise fooled us into believing art should be free. But it's also clear that we're again in the grip of a system we do not understand and cannot help but trust (the digital art market) and that we're outsourcing our choices to our emotions (the "feeling" Blau says people are after) instead of to our rationality (which would otherwise tell us that digital art is interchangeable with other copies of it you can easily find on the web). Against that backdrop of artificial scarcity (hello, Sendhil Mullainathan) and unconscious decision-making, fast-forward a generation or two. Will an artist make her own work, or will she only be remixing images from the past? Will an audience find that work, or wait for it to be recommended? And will we feel new emotions in response to new work, or will we feel only the emotions that The Loop spotted and rewarded years ago?

This sort of early prosthetic taste-shaping is taking place throughout the arts. A Los Angeles–based composer, Lucas Cantor, "finished" Franz Schubert's Unfinished Symphony by feeding the rest of the composer's work into a GAN and asking it to kick out its best guess at what the missing portion might sound like. (And it didn't take much to do it—all the processing took place on a single Huawei phone, as a promotion for the company.) "Everything that it sent me back I found to be kind of incredible," Cantor told me. "It's like having a collaborator that has an endless stream of ideas, that never gets tired, that never runs out of ideas, that never has a bad attitude, and never needs to take a break." By virtue of its mathematical structure, music is, in fact, one of the simplest human domains for AI to simulate. Services like Amper, Google's Magenta, and Flow of Machines only need a few suggestions from a human as to key, genre, mood, beats per minute, and then almost instantly produce a backing track you could plausibly hear in a movie or behind a rising artist. OpenAI's Jukebox even includes human singing that to my ear is indistinguishable from the real thing.

This uncanny simulation of real music isn't being done just to make us happier, of course. The larger purpose here is profit. A studio musician who belongs to the American Federation of Musicians charges at least $240 per recording session. Composer royalties, studio time, a recording engineer, the mix—all of it can drive a single song's price into the tens or hundreds of thousands of dollars. Amper offers a complete song to use in an online advertising campaign for only $499. Put aside the aesthetics for a moment. Consider what capitalism will do with this. Imagine just how quickly companies will want to seize on AI to individually tailor never-ending, never-repeating, low-cost entertainment for each of us. As Miles Brundage, a researcher at OpenAI wrote in 2020, "It seems safe to say that the era of almost-exclusively-human-generated and almost-never-individually-customized media will not last much longer."[4]

It may not be that *any* of this is art, in the philosophical sense. It is, of course, all just an imitation, reverse engineered from the echo of an audience's reaction to past art. But if you or I can't tell the difference, what does it matter? Our minds will come to expect it. And soon we won't know any other way to live. Studio musicians and original composers and poets and painters will find themselves in a world where art is algorithmically matched to a few of our most basic emotions, where its creation is push-button easy, and where true surprises are almost impossible to come by.

NOW LET'S ADD another layer of complication, one that's an integral part of what we're discussing, and that's the way that capitalism blinds us to the inequities baked into The Loop.

We've always believed that hard work and good fortune are the ingredients of a successful life. American capitalism, and the English agrarian individualism that helped spawn it, were born out of the idea that we work hard, and perhaps our innovations will more efficiently harvest the crop or pump the oil, but the chance for anyone to get lucky is also just part of what we consider proper functioning of the market. Bump into the undervalued home, the singer making a comeback, the overlooked oil lease, sign at the right moment, and maybe you, too, can strike it rich. There's room for chance, novelty, spontaneity.

But by sampling what has come before, AI could fundamentally do away with all of that even as we continue to believe that the possibility of a lucky break is the great leveling force of capitalism.

Of course, that belief that anyone can make it has always been a flawed one, because as has been demonstrated again and again, hard work and good fortune cannot overcome centuries of systemic bias when it comes to getting a home, a loan, a job. The narrative of a lucky entrepreneur with a good idea really only fits those lucky enough to have been born into the right race and gender and last name. So it's frightening to consider how The Loop might continue to keep that anyone-can-be-a-star narrative aloft even as it makes that narrative impossible for more and more people.

Jesus Hernandez is an urban sociologist who spent thirty years evaluating welfare and mental health programs for the state of California as a policy analyst while keeping his family afloat as a real estate agent in Sacramento. He was in his third decade going back and forth between crisis planning and open houses when the subprime crisis hit. He watched foreclosures in Sacramento rise from a record low of only 117 in 2005, the peak of subprime lending, to 17,801 in 2008. But the misfortune wasn't evenly distributed. While the crisis was more moderate in primarily white neighborhoods, homes in nonwhite neighborhoods like Oak Park and Del Paso Heights, which sit right by the downtown business district, lost as much as 80 percent of their value from their 2006 peak.

The experience was so disturbing that when Hernandez took his state retirement, he put the money toward getting a PhD in sociology at the University of California at Davis and became an expert in the mechanism of risk in housing, as well as how it becomes corrupted. "There are four preconditions for a functioning market," he told me. "Buyers and sellers, property rights, information, access to capital. When you interrupt those conditions, that's when you change how the market performs." Again and again, he says, the patterns of interruption are easy to spot, all over the nation. "When you look at housing in the US, you can see how we've skewed those conditions in multiple ways."

Hernandez began to see such concrete patterns that he spent three years of his own time and money to assemble data and the academic literature for a citywide analysis of Sacramento that would predict, down to a single household, how conditions can support or destroy a chance for stable housing, and for a successful life in general. "My thing was 'what happens to a neighborhood when it gets no love from a city?'" he says. His map divided Sacramento into quarters, and he found that the north-south axis was mostly poor and nonwhite, while the east-west axis was comparatively wealthy and mostly white. Those patterns perfectly matched the historical geography of redlining—the places in Sacramento where so-called race covenants written into the paperwork of homeownership forbade selling a property to a nonwhite buyer. When Hernandez overlays that exclusionary history on a map with the federal and state funding for better housing, he says, "I can predict where gang violence happens, where transit dependence happens, who's going to lose their schools, all of it."

Based on that map, he has now built a system for forecasting a neighborhood's trajectory depending on which necessary conditions for health and wealth have fallen away. "When one of those is missing, the rest of it goes bad. Housing wipes out education, which wipes out your job, and so you're not going to have any health benefits, which screws up your housing," he explains. Other cities are now using his tool to evaluate themselves.

Between his two careers—one spent trying to understand how people become trapped in the most desperate circumstances, the other spent trying to help people find their way into a new financial life—Hernandez says he now understands that the grand circus of opportunity isn't open to all Americans, that vast numbers of people have been handicapped by more than a century without the proper opportunity and investment, and that the "invisible hand" described by Adam Smith as equalizing the open market doesn't exist.

"If we could see the invisible hand," he told me, "we'd see that it's white." As he explains it, when a neighborhood has been held in stasis for as long as Sacramento's nonwhite neighborhoods have, it begins to look as if the conditions are somehow endemic. We as humans fall prey to the illusion that "bad" and "good" neighborhoods are a reflection of the moral standing and ongoing creditworthiness of their occupants, rather than a reflection of the discriminatory practices that created impossible conditions for them. Now imagine feeding the data on crime, home loans, house fires, and alcoholism across Sacramento into a database. Imagine putting a pattern-recognition algorithm to work choosing which Sacramento residents are most likely to repay a loan or die before they reach American life expectancy. The algorithms won't tell us *why* that might be the case. They'll only tell an insurance adjuster, a bail bondsman, or a mortgage broker what the statistics have been in the past. And in the process those algorithms will absolve those humans of the legal or moral responsibility to look into the root causes of the inequity, the poor health, the high default rates. Here, the algorithm says. Here's where history says you should invest. This is The Loop following the same choreography: the data is sampled, the results are analyzed, a shrunken list of choices is offered, and we choose again, continuing the cycle. But when you see that the same systems that determine which authors you might enjoy (and shape the market such that similar authors become dominant) might also determine creditworthiness (and the shape of the housing market), the stakes of The Loop suddenly become a whole lot higher than just what music we're listening to.

There's nothing more inherently, blindly discriminatory than pattern-recognition algorithms. To our anthropomorphizing instincts, they look like nothing more than neutral arbitrators. And they get the humans that use them out of sticky moral and legal dilemmas, the kind of decisions we love to offload. How will we fight all of that even as we hand algorithms more and more decision-making power?

BIAS AND DISCRIMINATION in machine learning is the earliest, ugliest of the many ethical questions involved in creating and deploying artificial intelligence. This is not to say that we're in any way prepared to solve those glaring problems, or that doing so will be easy. It's not even clear that if we clearly articulate the problem and outline a solution, people in a position of power will be willing to act on any of it.

But there is at least some sort of legal framework for labeling and addressing discrimination. When it comes to the central systems of modern life—employment, housing, criminal justice, education—discrimination on the basis of something like gender, race, or sexual orientation is against the law. And when we encounter pattern-recognition systems that turn out to amplify that prejudice, we can start from an agreed understanding that prejudice is bad and illegal.

The difficulty, of course, is that pattern-recognition systems are difficult to examine for something like discriminatory choices. Machine learning happens inside a black box of improvisational computational effort—literally a roomful of monkeys assigned to peck away at typewriters in the hopes of eventually and accidentally writing Shakespeare. At the moment, and because of the nature of how these systems are built, it's largely impossible to go back and ask the monkeys how they came to type out *King Lear.*

When I speak with data scientists about inequality in lending, they tend to have one of two reactions. One is that trying to compensate for past inequalities will ruin companies, and it is somehow morally or intellectually wrong. "What every lender in the history of the US has ever found is that pretty much any credit model will give people of different races different treatment," one builder of such models told me. "If you really wanted people from every race to be approved at the same rate, you'd really have to put your finger on the scale in a big way," he argued. "You can't build a business that way, unless it's hugely subsidized." The other reaction is that the system just needs to excise bad input. There's obviously something in there skewing the results, they point out. You just have to find the offending data in the algorithm's training set and remove it.

But what many bias researchers and technologists increasingly argue is that there is a pattern of systemic racism and institutional discrimination that companies have absorbed, learned from, and regurgitated in a way that no one, not even they, can specifically detect. And if we can agree that we should fight that pattern off, we can't *remove* data from the process. We have to instead stop pretending that numerical values and historical statistics are all the data we need. We instead need to account for past patterns of discrimination, the kind of horrific systemic abuses Jesus Hernandez has spent decades measuring, and in fact put our finger on the scale to compensate for it where we can.

There is vital and rapid work being done on bias in AI. The researchers Joy Buolamwini and Timnit Gebru published a seminal paper in February 2018 revealing that the top three commercial facial-recognition systems misidentified white, male faces only 0.8 percent of the time, while the same systems misidentified dark-skinned women more than 20 percent of the time. And the stakes, they pointed out, are high: "While face recognition software by itself should not be trained to determine the fate of an individual in the criminal justice system, it is very likely that such software is used to identify suspects."[5] Inspired in part by that work, a study by the federal agency charged with establishing technical benchmarks on new technology, the National Institute of Standards and Technology (NIST), found that across 189 facial-recognition algorithms from 99 developers around the world, Asian and African American faces were misidentified far more often than white faces. But NIST also found that this isn't because nonwhite faces are somehow harder to identify. In fact, the discrepancy between white and Asian faces disappeared entirely when the software was made in Asia.

Since NIST's study, commercial systems have improved their accuracy. Peter Trepp, the former CEO of FaceFirst, a facial-recognition provider, told me that since 2018 his company has achieved near-perfect accuracy across every skin tone, gender identity, or other personal variable. In thousands of stores across the country, facial-recognition systems quietly compare your face to a database of known shoplifters. I've even visited stores where a facial-recognition system tells the clerk my name and my shopping history, allowing him to approach me and see whether I'm interested in another suit like the last one I bought. Those are the kinds of systems FaceFirst makes. "We're seeing 99.8% accuracy, because we're seeing very large data sets that you can test publicly, that have a whole array of folks in them, from all different kinds of backgrounds," Trepp says. "So the claims about missing people of color, and some of the things we saw in years past, while they keep coming up in years-old media reports, they are really largely gone today. And that is thanks to making sure we train the data properly." As he describes it, the companies that buy software like FaceFirst to spot repeat shoplifters and regular customers aren't interested in technology that can only identify white men. They need something that can identify everyone, and the industry is highly motivated to provide that.

But sorting out the math of identifying faces accurately regardless of race is only a solution to a small part of the many problems I worry The Loop may impose on us. At times, as I learn about companies developing technologies to identify the emotions on our faces, to use our facial features to better target ads at us, to track our movements on a map by the various times we've shown up in a camera, I've wondered whether it might not be *better* to be misidentified much of the time. This is wrong, of course. As Robert Williams can tell you, misidentification doesn't just leave you out of systems you don't want to be part of; it can also get you arrested by them. But reliable accuracy in these systems isn't necessarily a cause for celebration. It just pushes us further along on The Loop toward a day when *everything* about us is being accurately evaluated and powerfully shaped.

And when I say everything, I really mean *everything*. It's tempting to believe that some aspects of human behavior can't be captured by pattern recognition, that there must be ineffable human attributes beyond algorithmic analysis. But I think that ultimately, with enough time and money and processing power applied to our behavior, along with the collapsing number of choices we make, as we get used to choosing only what's been offered to us, it will be possible to generate reliable-enough rules that we use algorithms for everything. We've shorthanded rules for seemingly immeasurable things before. In 1964, the *New Republic* writer Albert Goldman wrote an article about Lindy's, a New York delicatessen where television comedians dissected each other's work and tried to predict who was going to make it. These "bald-headed, cigar-chomping know-it-alls," Goldman wrote, had come to believe that "the life expectancy of a television comedian is proportional to the total amount of his exposure on the medium." Star in a weekly series and you're done for. Keep it to specials and guest appearances, and you might work into your golden years. Goldman titled his article "Lindy's Law," and successive generations of writers seized on his theme as a way of predicting the longevity of certain ideas and works of art.

This wasn't just wannabe *Late Night* writers ruminating over bad coffee. Goldman's concept launched a whole world of statistical thinking. The mathematician Benoit Mandelbrot updated the model in a 1982 book about fractals to posit that comedians who have made appearances in the past are more likely to make them in the future. Nicholas Taleb carried Mandelbrot's concept into his book *Black Swan*, and in his book *Antifragile* put a specific math to the notion that as an idea survives, its longevity *increases*. "Every year that passes without extinction doubles the additional life expectancy," Taleb writes.

In a sense, Goldman had identified the kind of law AI might use to evaluate the prospects of a comedian. This lineage of ideas suggests it's very possible that human taste, and the market forces that seek to satisfy it, are bound by rules that truly can be expressed and perhaps even used to make predictions. Of course, Goldman wrote in 1964 that "Lindy's Law, despite its awesome air of common sense, is more a cautionary fable than an accurate description or prognostic." The factors that determine a successful career in comedy or in television are too capricious and complex to be predicted by a simple law.

But he was sure of one thing. Writing of the offbeat comic genius Jonathan Winters, whose trajectory in television had been the subject of much debate at Lindy's, Goldman worried that television executives, "frightened by the unpredictability of the comic's mental fugues, will urge him to develop a safe, clean, dependable formula." Business likes formulas, and where it can't *predict* behavior, it wants to *shape* behavior.

Today, we're much more likely to be able to develop a system that truly *can* predict which comics are going to make it. And by doing so, we'll be flattening out what makes individuals special in a landscape of statistics, leaving behind writers too weird for Wattpad and the people from the wrong part of Sacramento, and rewarding bland, affable TikTok stars, artists who make art like other popular artists, and the dullest possible version of Jonathan Winters.

By using AI to make choices for us, we will wind up reprogramming our brains and our society. Leaning on AI to choose and even make art, or music, or comedy will wind up shaping our taste in it, just as it will shape social policy, where we live, the jobs we get.

A February 2020 UC Berkeley study pointed out that supervised learning tends to make predictions that wind up influencing the thing being predicted, a tendency called "performativity":

> Traffic predictions influence traffic patterns, crime location prediction influences police allocations that may deter crime, recommendations shape preferences and thus consumption, stock price prediction determines trading activity and hence prices.[6]

Until now, the authors note, performativity, which is well-understood in academic and policy circles, has been largely ignored by AI engineers, even though it comes up all the time. "When ignored," the researchers write, "performativity surfaces as undesirable distribution shift." The landscape changes as it's being analyzed, but it looks like a data problem, and so the best that AI creators have thought to do is keep retraining the AI based on past outcomes, just as the artist Mario Klingemann nudged his AI-driven portrait machine this way and that in pursuit of something he liked.

To avoid this, the study concludes, perhaps AI shouldn't evaluate its predictions against past outcomes, "but against the future outcomes that manifest from acting on the prediction." They go on to lay out a possible mathematical framework "in which an agent first collects a batch of data under a particular policy . . . and then finds the optimal policy for that trajectory *offline*." Rather than choose a screenplay or construct a piece of art or music based on what we've liked in the past, perhaps it will be necessary to tweak the AI's instructions "offline"—to jam a fist into the pudding and scramble it a little. Or at the very least, before we ask AI to entertain us, choose job candidates, or approve home loans, we need to decide for ourselves what we *want* for one another in the future and push it to deliver *that*.

But in the meantime, it will be important to remember that just because AI isn't delivering the results we need or want doesn't mean we won't be convinced by them, or that we won't be subjected to them again. Once AI-curated screenplays are greenlit and AI-generated art is sold, that's the data being sampled for the next go-round of The Loop, sending us down a path of narrowing options.

This will also be true of AI that claims to predict things we'd rather a machine judge for us anyway. It's our greatest success as a society to have devised systems for settling our differences, measuring the immeasurable, sorting through controversy. It's not easy, and the work isn't finished. Mediation, negotiation, deciding which of two bad choices is less harmful—that's the very hardest stuff of modern society. In some cases, we've barely begun to tackle the challenge. But now that a technology has come along that can, in theory, take over some of that decision-making for us—a technology that forecasts outcomes based on the past—we're going to begin handing important choices to it, because they're choices we don't want to make.

YACQUELINE AND I are sitting together in a kitchen in the suburbs of Northern California, and she's explaining how she decided to dismantle her relationship with her son's father. (I'm withholding her last name at her request.) They were together for five years before she realized something had to change, and she spent another two years searching for equilibrium. "We tried everything. We tried counseling together. We tried going to church." But something just stood in the way. "I just wanted us to be able to be friends," she says. "How we were before the baby. Before everything happened. But since we had so much history we couldn't."

For generations, the prevailing assumption has been that conflict between parents is so terrible for the child that if a couple cannot get along, one of the parents should take sole custody. But dozens of studies[7] in the last twenty years[8] have found that children benefit from joint-custody arrangements even if the parents are in conflict much of the time, and that winner-take-all custody arrangements not only rob the child of time with a father or mother who loves them, but can also intensify conflict between the parents. To me, separation and divorce look too complicated to quantify in any way that AI, much less human researchers, could analyze—a vast, immeasurable storm cloud of individualized triggers, childhood traumas, flung phones, and raised voices. But it turns out that conflict between lovers, and especially between parents, follows extremely predictable patterns. In fact, statisticians have found the individualized factors to be so consistent, they describe the three big ones in a single acronym: PIC, which stands for the quality of the parent-child relationship, the family's income level, and the degree of conflict. And when you control for the PIC factors, it turns out that joint-custody arrangements work out better for a child than putting her or him in a sole-custody arrangement.

The legal and social model for parenting as a divorced couple seems to be adapting to this new understanding. In Western nations, and particularly in Scandinavia, where the most research has been done, joint custody is increasingly common. In the United States,

states like Wisconsin see as many as 30 percent of children in custody cases going into joint arrangements. In Sweden, it's over 60 percent.

But for parents like Yacqueline, speaking to me through the fatigue of managing a full-time job, a young child, and a man with whom she cannot get along, the job of putting all of that together was too much to do on their own.

"I felt like we were both adults and we should both be able to talk things out and put our emotions aside to raise our child together. I figure that would be the logical thing to do. And we tried that for a couple months." But she and her ex simply could not function together, she tells me. Even the smallest logistical negotiations would devolve into verbal abuse. "So finally we went back to court," she says. The judge, looking at their history of mediation, co-parenting classes, even a temporary restraining order, told them that from now on, all of their communications had to go through a new, AI-fueled mediation system.

Yacqueline reaches into her purse and brings out her phone to show the system to me: a communication and recordkeeping app called coParenter. She swipes through a series of screens, shows me text messages between her and her ex, a calendar of doctor appointments and karate lessons, and locations on a map where she or her ex has indicated they picked up or dropped off their son. She also shows me the times in text conversations when the software has interrupted communication between Yacqueline and her ex with a pop-up message, urging them to think of their child, or offering help. It gives her the chance to speak with a mediator, for instance, and asks whether she'd like to speak alone or include her ex. Then it offers a list of likely topics to sort through, from communication to financial disagreements to daycare.

coParenter locks onto the patterns in how couples tend to negotiate with one another, and uses machine learning to anticipate what's coming next in the conversation. It turns out that stuff is actually pretty easy to spot. "When people want to swap days, or get extra days, there are only so many ways to ask for that," says Jonathan Verk, who cofounded the company. His system uses AI to look for patterns in what couples ask of each other, and to then recommend an arrangement the moment the language in a text exchange suggests that this negotiation is like others the system has seen. "We have 20,000 agreements that have come together about swapping days, trading holidays, that kind of thing," he says. When Yacqueline and her ex began talking about karate lessons for their son, the system interrupted their chat with a perky suggestion that they simply adapt an informal agreement that had been used thousands of times before by other co-parents sending a child to everything from summer camp to cello lessons, and they wound up adapting it as an agreement between them. And it worked.

Where the app truly tips us all toward the future is where it not only tries to forge agreement, but actively attempts to head off disagreement. Yacqueline, for my benefit, carefully types in a message full of foul language, and up pops a warning politely explaining that this sort of abusive language can set everyone back. "I have a potty mouth, but I don't say as many bad words anymore," she says with an embarrassed laugh. "It's like bowling! It sets bumpers at least so you're going in the right direction."

Yacqueline says she loves this sort of guidance, that it has only improved the way she and her ex behave toward one another. co-Parenter claims that more than 80 percent of the couples who begin using their software never go before a family court judge again. That's been true for Yacqueline. She says the program has been transformative, and being supervised by software in no way bothers her.

"My son's only five, he'll be six soon. So we have a good amount of time until we're done dealing with each other." I ask Yacqueline if she can imagine using coParenter all the way until her son is an adult. "Oh absolutely," she tells me. "I use it on a daily basis. I don't see myself communicating outside of that."

Yacqueline says she can detect that there is a larger pattern at work, that in many ways our relationships are an echo of the prior generations' relationships, and that for her and her ex, as for any couple, breaking out of that loop is very difficult.

"My mom and my dad weren't together. His parents weren't together, either. So we grew up in a pattern of not having both of our parents. So we don't know what it's like to have both of them." She sighs. "The way he reacts and the way I react: it's similar to what we've been through." Like any parent, Yacqueline is making it up as she goes along, and she says that being coached along the way—I think of Niec speaking into Heather's ear, guiding her through interactions with her son Miles—is incredibly helpful.

What is uncharted here is whether Yacqueline and her ex are going to be creating a new echo for their son, one that will affect him long after they're no longer in charge of him. This is The Loop reading our behavioral patterns, and then steering our behavior for us. Systems like coParenter, which find the patterns we cannot see ourselves, ostensibly stand in for our System 2, the higher functions that have helped us build systems of law, mediation, morality. But they only engage our System 1. With coParenter at work, we don't have to take a breath and think about the bigger picture. We can simply follow the onscreen prompts more or less unconsciously.

I don't know whether this is good or bad; I only know that the effect is incredibly powerful, and it's likely to last across generations. Yacqueline's son is receiving the benefits of being able to enjoy both of his parents. He is watching them somehow get past their tensions and function together through the mysterious mechanism hidden inside their phones. But what is he learning? Is he learning *how* to communicate with the most difficult people in his life? Are Yacqueline and her ex modeling anything for him at all, except the bland collegiality they're reading line for line from a series of AI-driven prompts? Will he grow up with an idea in mind as to how separated parents, rival coworkers, estranged friends should communicate? Will that be in his brain somehow? Or will he just come to rely on what will, by then, be a sophisticated but prosthetic set of social instincts, communicated by app? Is this training? Or are these training wheels that never come off? Will The Loop cause successive generations to lose sight of what it is they like about art, about music, about each other?

BY NOW I hope it's clear that coParenter is part of the final loop I set out to explore in this book. It belongs to a new wave of businesses deploying endless products that claim to not just spot patterns in human behavior, but also shape them. Those businesses are growing rapidly.

Boston-based meQuilibrium uses AI to analyze the digital communications of each employee in a company to form "a baseline resilience score, profile and tailored plan to address their areas of greatest vulnerability," according to the company's online marketing materials. Basically, it reads their communications and makes automated suggestions about how they can improve as individuals and how the company can better manage all its employees overall.

Cogito, another such company, is a sort of real-time coach for call-center agents. Its software analyzes phone conversations with customers and looks for signs of frustration, hurried speech, even "intent to buy." As the call-center agent progresses, he might be given an "empathy cue" if the software thinks the conversation needs a bit of warmth. According to the company's sales materials, not only can agents use the technology to "speak more confidently, concisely, and compassionately," but the technology also looks for the common characteristics of the highest-performing agents, and "provides automated guidance to help all agents perform like the best." And by looking for patterns like mimicry, consistency, turn-taking, "a live dashboard shows the behaviors at which agents are excelling, or highlights those which require attention." Cogito's cofounder and CEO, Josh Feast, says the results are so promising he's interested in expanding the technology to other social areas, like communication between coworkers. (Writing in *Fast Company*, UNC Chapel Hill professor of law Jeffrey Hirsch points out that "except for some narrow exceptions—like in bathrooms and other specific areas where workers can expect relative privacy—private-sector employees have virtually no way, nor any legal right, to opt out of this sort of monitoring. They may not even be informed that it is occurring."[9])

And then there's Affectiva. The company's CEO, Rana el Kaliouby, first put AI to work in an attempt to help communication between people with autism and the rest of the world. But soon she began applying the technology to a broader and more commercial set of purposes, like scanning the driver of a car for signs of danger. El Kaliouby explained to me that her AI knows, for instance, that when the eyes are tired but the mouth has stopped yawning, a driver is likely about to nod off at the wheel. However, the company's newest foray is into true emotional measurement, such as attempting to discern whether someone has had a positive reaction to an advertisement or to a movie's pivotal scene. We know, thanks to the work of people like Paul Slovic and Wendy Wood, that we offload huge amounts of our decision-making to our emotions, that "going with our gut" is one of our primary behavioral mechanisms, and that we have almost no conscious control over any of it. Giving AI a shot at measuring our emotions is a vast business opportunity, and it paves the way toward influencing our emotions and the decisions we make with them.

Many critics have argued that these technologies simply can't do what their creators claim. UCSF professor of psychiatry Wendy Mendes, who has made a career studying the ways our emotions shape our behavior, says that the cultural and individual differences between people when it comes to how our faces convey anger, or humor, or sadness, make it nearly impossible, not to mention dangerous, to try to read them through a trained algorithm.

"Am I angry, or do I want you to think I'm angry?" she asked me during a lunch together in San Francisco. Microexpressions, she told me—the tiny movements of the face that AI can be trained on—are just too subtle and too varied for some sort of standardized, automated analysis. "This belief that microexpressions are a read-out of an inner state is a problem. They're for communicating an inner state, and they're culturally specific," she says. "Taking this idea that there's this perfect, systematic, statistical regularity simply doesn't hold up."

And beyond that, she believes that the promised convenience of a system that claims to be able to detect intent—the intent to steal or to swing a fist, for instance—will be too tempting for a police officer or federal investigator. "Law enforcement agencies will make reverse inferences if this stuff becomes widely available," she tells me, because the possible time-saving potential is irresistible, even though as scientists "we beat each other up all the time in a paper when we try to make this reverse inference."

She points out, for instance, the SPOT program, a nearly billion-dollar effort by the Transportation Safety Administration based on the work of the American psychologist Paul Ekman. Ekman, who was a pioneer in the study of microexpressions, helped TSA create a system that relied on a taxonomy of facial expressions. "When there's a lot to lose, such as death or imprisonment, the emotions generated are intense and very hard to conceal," Ekman wrote in a 2015 blog post defending the program, "so they often leak out."

"I actually have the endowed chair in his name," Mendes told me, "and I have huge respect for him. But taking his work and assuming that the basic emotions are invariant across culture? That just hasn't held up."

In November 2013, the Government Accountability Office released a report that found that after spending nearly $900 million on the program, the TSA had no evidence it worked whatsoever. That day, then South Carolina representative Mark Sanford grilled the head of the TSA at the time, John S. Pistole, rattling through the potential reasons the SPOT program might pull someone out of line.

SANFORD: If you were a staunch right-wing conspiracist with very strong anti-government leanings, you've posted some things that probably weren't the best to post on the Internet, but you had the invisibility that goes with the Internet, but now you've got a law enforcement officer probing, asking you questions, would you exhibit stress or fear?

PISTOLE: It depends on the individual, but potentially, sure.

SANFORD: You're an immigrant whose dad and mom came here illegally, would you exhibit stress or fear if someone was asking you questions?

PISTOLE: All situational, again.

SANFORD: Let's say you're a wife whose husband had beaten her and you're just trying to get on an airplane and get out of town, would you exhibit stress or fear if someone was going into interrogation on some front?

PISTOLE: Again, situational.

SANFORD: Which I think raises the point which the GAO report has brought. You go through a screening system which essentially undresses somebody, you send their equipment through radar detection and other devices. The question is, from a civil liberties standpoint, given those other tests, do you in addition have to go through a screening process based on somebody's interpretation of what might be in your brain?

PISTOLE: You raise good points, congressman. There's no perfect science, there's no perfect art of this.

But in the promotional materials of the companies selling AI-powered behavior-analysis systems, the science is made to sound perfect. And while that's undoubtedly not yet true, the patterns that AI *can* spot look credible enough and, as Mendes pointed out to me, are *convenient* enough that we'll be sorely tempted to adopt them wholesale.

Already, you can see it happening. In 2017, researchers at Carnegie Mellon University published code on GitHub called OpenPose, described as a "real-time multi-person keypoint detection library for body, face, hands, and foot estimation." A year later, a startup called Earth Eyes, together with Japan's telecom monolith NTT East, turned it into a system called AI Guardsman that scanned people passing through a store for signs they might be about to shoplift something, and alerted store owners in time for them to intervene. And by 2019, more than fifty stores in Japan had signed up for a competitor, Vaakeye, which similarly promised to spot "restlessness," "fidgeting," and other behaviors it says indicate possible criminal intent. The Loop is growing and growing.

It's also beginning to affect what we know about the world around us. Once upon a time the creation and placement of advertising was a human affair, in which creative teams took a flyer on a campaign idea, got it past a client, and then an account-management team placed it in the newspapers, magazines, radio, and television stations thought most appropriate to find the right audience.

In 1994, *HotWired*, then the online arm of *Wired* magazine, sold the first Internet banner ads to AT&T, in the typical choreography of phone calls and celebratory dinners that had until then put an estimated five hundred analog ads in front of each American per day. But today, programmatic advertising, which automatically places ads across billions of pages, videos, and news feeds, means the average American sees something closer to 10,000 ads a day.

The sheer scale of all that matchmaking has essentially wiped out human involvement in what was once called account management. And the resulting industry has produced huge amounts of money. By 2016, online advertising revenue had shot past that of television, once the dominant advertising medium.

But now that online advertising is a lucrative, inextricable aspect of our daily lives, a few hidden secondary effects are coming to light. It's already well known that the rise of online advertising and the online news outlets it ostensibly supports has damaged the things print journalism once did best: drawing readers in, keeping them loyal to a product, and engaging their attention long enough to truly inform them. Local newspapers have all but disappeared, unable to earn adequate advertising revenue off their small regional audiences. The rise of social media has essentially wiped out the opportunity for the publisher to make money—Facebook earns money on the ads it serves alongside journalism in your news feed, the publisher does not. (Platform companies like Facebook and Google argue that they deliver money to publishers by directing traffic to their sites, and that publishers choose to post their stories on social media because doing so brings them value.) And as more and more intervening layers come between the reader and the original source, not to mention so many other diversions competing for our attention online, our attention span dissipates. A 2020 Pew Center study found that people who rely on social media for news not only don't follow topics like the coronavirus or a presidential election as closely as people who engage directly with news sites or television news, they also know less about the subjects.

Meanwhile, advertisers who could have once picked up the phone and asked their agency to make sure not to place their ads in, say, pornographic magazines, or inside a radio broadcast about overthrowing the government, now rely on an industry of companies that make what's known as "brand-safety technology." These are companies that deploy keyword searches and AI analysis to blacklist certain sites for the sprawling brokerage networks that place ads in your path across the web. If the tech identifies a page as containing something gross or dismaying, that brokerage won't put an ad there.

Together, these two forces—the disappearing financial viability of journalism, and the growth of automated advertising placement—have given rise to another circle of The Loop.

A computational biologist named Krzysztof Franaszek decided after grad school to apply skills he'd once used to analyze data about Ebola and cancer to the world of advertising technology. And in 2020 he published the results of a shocking study into the ways that brand-safe technology was determining the flow of money to online publishers. In essence, he discovered that The Loop is defunding the news.

Franaszek wrote that "an estimated 21% of *economist.com* articles, 30.3% of *nytimes.com*, 43% of *wsj.com*, and 52.8% of articles on *vice.com* are being labeled as 'brand unsafe.'"[10] And when he drilled into the individual reporting of particular correspondents, Franaszek found that editorially invaluable beats were particular black holes when it came to online advertising. The work of Ali Winston, a *New York Times* reporter covering crime and surveillance in the United States, was labeled "unsafe" for advertising roughly 96 percent of the time. In other words, stories by Winston were not being advertised against, making them less directly valuable to the *Times*. This sort of inadvertent algorithmic blacklisting was happening to Pulitzer finalists and was making coverage of some of the most important stories of the moment a literal money-loser for news outlets. Just think of what these algorithms might make of the Jeffrey Epstein trial, of death in Gaza, of morgues overflowing with coronavirus victims.

Nandini Jammi, who runs a brand-safety consultancy for advertisers, explained the problem to me this way: "Instead of actually looking for criminal content, these brand-safety technology vendors have identified keywords such as 'crime,' or 'violence,' or 'terrorist.' So they're literally blocking words that journalists need to use to cover the news." It's a drastic simplification of our online behavior as data—the kind of thing a brutally efficient algorithm given an objective outcome will always do—and the most business-minded possible response to it.

It also doesn't account for the subtleties of media literacy. As Franaszek put it, "Brand safety tech appears to be highly sensitive to the presence of certain keywords, but cannot assess publisher context, image content, or misinformation."

Jammi summed up this lucrative, ostensibly sophisticated technology neatly: "The way it works is stupider than you can possibly imagine."

COMPANIES ARE DEPLOYING pattern-recognition systems to guide us in love, work, risk, justice, and it's clear by now that the effects go far beyond the simple efficiencies they offer. It's not just the stuff of short-term convenience, like a list of suggested movies you might enjoy tonight. The same algorithms that make those sorts of recommendations are stepping into corners of our life we thought were too, well, *human* to be digitized. But it turns out that just as we'd rather have a computer decide which four people are getting kicked off a flight, we'd rather have a computer decide all sorts of other complex interpersonal and moral stuff for us as well.

What will the long-term effects be? Will we wind up modulating our behavior in stores to avoid being accused of planning to shoplift?

Will some people set off these systems inadvertently? What about people of color who are disproportionately surveilled by store personnel all their lives? Will they seem restless and fidgety, and inadvertently become part of a loop of scrutiny into which other people of color will continue to fall?

We will wind up using AI on the stuff we'd rather not do—morally and legally fraught choices about who gets a loan or a job, risky gambles on screenplays and advertising. And if we succumb to the idea that we can pay a company to provide decision technology that absolves us not only of the difficult experience of making these judgments but of any culpability for the biases we might hold about one another, before we know it we won't know how to do it any other way. Studios won't pay for film scripts that haven't been endorsed by AI based on past hits. Companies won't have any way to make an exception to their AI package's judgment when it comes to whom they hire, and legally, they won't be able to. Mortgage brokers will be hostage to machine-learning-based evaluations, which in turn will reinforce another hundred years of inequity. Unless we begin developing tactics for resisting all of this, The Loop will eventually collapse our options in any area of life we let it sample.

What happens then? At that stage, the problem will no longer be that the technology is inaccurate. It's that we'll have gotten entirely used to relying on it, the way Yacqueline and her ex rely on coParenter to make difficult decisions about one another. And if that's the case, we'll have essentially reprogrammed not just the systems of our lives. We'll have reprogrammed *ourselves.*

Chapter 10

MISSION CRITICAL

I ONCE SAT inside a flight simulator, slowly blacking out.

"Flight simulator" doesn't do the machine justice, but those are the right words. There are the *software* simulators, which show you the ride. And there are the physical simulators, which bump you around on a small crane. This wasn't like that. This was the whole experience, and I was about to wash out.

I was in a windowless white capsule roughly large enough to replicate the inside of a fighter jet. The capsule was attached on its right side to an enormous joint that could rotate it in any direction. And the whole thing sat suspended at one end of a roughly one-hundred-foot-long centrifuge that filled a hangar-sized room.

The contraption can produce pretty much any sensation an airplane can—certainly as much as any non-pilot would ever be conscious enough to experience. The trick is that while I was actually spinning in a circle, the image on the screen fooled my brain into thinking I was traveling a linear path through open sky. I had already made it through what the manufacturers consider a modest test of acceleration. As the screen image told me I was rocketing upward toward an imaginary sun, the centrifuge spun faster and faster, and eventually I experienced six times the force of gravity, enough to crush my face and torso into the seat. It was scary and deeply uncomfortable. My Adam's apple, which normally protrudes from the front of my throat, pressed against the back of it as the machine whined up to higher and higher speed. But it turns out that this sort of g-force—along the front-to-back axis of the body, the kind of force that produces gruesome photographs of handsome pilots contorted into goblins—is the easiest to handle. It's when the simulator gives the experience of downward Gs, which compress the body toward the feet, that things go really wrong.

As the capsule did roughly half a barrel roll, so that my feet were now on the very outside of the centrifuge's rotation, imitating the sensation of hard banking, I tried to do what they'd told me: squeeze my butt, press my feet down on the pedals, all in an effort to keep blood above my waist. But it was no use. At 3.5 Gs, my brain was losing circulation. Colors appeared, then a tunnel of black slowly formed at the edge of my vision, and pretty soon I was blind. "I can't see," I gasped, and the operator slowed the machine just before I lost consciousness.

It took me roughly an hour in a folding chair to recover. Not only was there the nearly passing out and all that; the conflict between what my inner ear was experiencing (rotation!) and what my eyes were being shown inside the capsule (level flight!) formed the classic internal miscommunication that researchers think creates motion sickness.

But eventually I got it together and sat down with a flight surgeon to find out why I'd fallen apart. He explained that the blacking out is an interruption of the hydrostatic loop—the liquid connection between your heart and brain that keeps you conscious. When the downward pull of those g-forces breaks that connection by pulling your blood toward your feet, that's it.

He also told me that his company sells this simulator to allied air forces all over the world, and that every so often those air forces will link up their simulators in one enormous virtual airspace, bring in their best pilots, and dogfight. It's like a high-stakes video game tournament, with barfing and unconsciousness and global bragging rights.

"Well?" I asked. He blinked at me. I laughed. "Who wins?"

"The Bahrainis," he said without hesitation.

This took me by surprise. Bahrain is a wealthy nation, sure, but it has a population of less than two million people. "Why? Why are they so good?"

"Well, it's not really that they're good," he said carefully. "It's that they're, uh, not very tall." He looked embarrassed. "Their hydrostatic loop is shortest, so it's hardest to break. They can turn tighter than the other teams."

It turned out that the system that determines the outcome of those battles isn't the weapons, or the flight surfaces, or the armored exterior. It's the human tendency to fall apart when our bodies move in the wrong way. In these virtual dogfighting contests, the difference between winning and losing (and with it, in theory, preserving or losing a $90 million combat aircraft) amounts to who among the pilots has a heart and brain located closest together.

Former secretary of the navy Ray Mabus famously said about the F-35 fighter jet, the newest single-pilot combat aircraft, that it "should be, and almost certainly will be, the last manned strike fighter aircraft the Department of the Navy will ever buy or fly." He later walked it back, but his instincts are understandable. The pilot is any plane's most valuable cargo, but also its greatest weakness.

This is why militaries all over the world are automating everything from armed sentry duty to anti-aircraft guns to cyberwarfare. Humans aren't up to the job. And yet that will call upon technology to make decisions that we humans can't agree on. And in the process, we're changing the way we make those decisions. Military technology is reprogramming our attitude toward war.

As we've seen, technology has already shaped our habits even when we're not trying to kill one another. When I land in a new city and need to drive from the airport into town, Google Maps offers me a handful of options—here's one highway, here's another, here's the no-tolls route. And I dutifully choose from among them. It used to be that I'd push back against this process and seek out whatever bakeries and landmarks and vistas I could find by driving unassisted. But these days I just go along with the choices. These navigation systems funnel our endless options into a multiple-choice question, and we wind up thinking there are only three ways into town. A navigation system feels as if it's giving us more options in our lives, but in fact we wind up with fewer of them.

So it is in military systems. The Boeing Corporation pioneered "mission planning" products that do to military decision-making what Google has done to your commute. "Mission planning" has been a product at Boeing since 1975, and now several Boeing competitors make them, too. In fact, a new DoD-wide project called the Global Information Dominance Experiment, announced in August 2021, is supposed to more or less predict the future using AI. "And I'm talking not minutes and hours. I'm talking days," General Glen D. VanHerck, commander of NORAD and US Northern Command, told reporters. These systems take all the complexity and urgency of military planning and do their best to reduce it all to a handful of choices. ("We don't have any machines making decisions," General VanHerck reassured a reporter. "Machines can provide options.")

Whether it's the need to airlift troops in or out of the field, plan a missile strike on an installation, or deploy soldiers to secure a town, the software funnels all the open-ended variables into a convenient menu. You might choose the scenic route from the list that Google offers you. A naval commander might choose the over-the-mountain flight path from the list that Boeing offers. We feed our choices into a system, it analyzes the outcomes, and offers a new menu of choices for us to choose from. Choose from that, analyze it again, and we're in The Loop.

But that's just logistics. Surely handing guns to robots isn't a thing yet? Sure it is. In 2013, at Fort Benning, Georgia, the army invited a handful of military robotics contractors to trot out their best attempt at a self-directed mobile weapons platform—robots with guns. It was one of the first times the United States offered any public display of their intention to automate lethal force. Various small tanks and wheeled boxes rolled to the firing line at an outdoor range. With operators standing around at a safe distance, giving each robot the final go-ahead to fire, they shot up a hillside full of targets.

That final go-ahead to fire is the number-one reassurance that military spokespeople offer when asked about the ethics of weaponizing a robot. The technical term is "human in the loop"—a human being will always, it is argued, be involved in deciding to use deadly force. But even if that's true, the role of the human is still very different when the robot is holding the gun. If one imagines a robot being sent into a hostile environment, or assigned to patrol a perimeter, the human will enter the loop very late. The robot will be able to detect an anomaly, designate it a threat, and put a gunsight on it. If anything, the forward momentum of all that preparation will reduce any human's role to little more than the role of veto. The robot makes all the moral and logistical preparations, and the human's choices are reduced to "shoot" or "don't shoot."

Humans are, in fact, already out of the lethal decision-making loop, because just as the organs of a human can't be trusted to keep them from fainting at the controls of a fighter jet, the senses and reaction times of a human aren't fast enough to handle the kinds of situations that combat now demands.

The SeaRAM system, built by Raytheon, is a last-ditch shipboard defense system for knocking incoming missiles out of the sky. A human would be essentially useless in this role, because the seconds between detection and impact simply aren't enough time to consult a human operator, much less ask her to track and shoot down incoming projectiles. So the ship defends itself with an automated battery of missiles that detect, track, and fire on incoming projectiles without any human involved. The company's tagline for the system is "evolved ship defense."

And in Israel, the Iron Dome system has been in operation for a decade, intercepting incoming rockets from Palestine. As early as 2014, smartphone video showed the system doing what used to be impossible. Two Qassam rockets—cheaply assembled pipes loaded with explosives and fitted with metal fins—are faintly visible, sputtering upward over Tel Aviv. In a moment, their propellant will give out, and they'll fall to earth. They have no guidance system; they're just straightforward ballistic weapons following a simple curve from launch to impact. They're nearing the top of their arc and could land anywhere in a city of more than three million people.

Then two flares brighten the rooftops, and a second pair of projectiles leaves the ground. These missiles behave very differently from the first two. They leave a bright afterimage as they fire upward over the low apartment buildings of the neighborhood, and they seem somehow alive, tracing a path in partnership. They level out and slow with short bursts of fire, hesitate, and then carve upward again. Suddenly, there is a pair of explosions as each of these self-guided missiles detonates. They have destroyed not only themselves, but also the Qassam rockets. The explosion dies away, the sirens wail on for a few minutes, and then everything is quiet.

In its first three years online, the system was credited with intercepting more than 1,200 rockets. In the first seven days of the 2021 fighting between Israel and Hamas, during its first use in open combat, it downed another 1,000, according to Israel, along with several drones.

Because the incoming rockets are typically fired from no more than roughly forty miles away, and at speeds of more than 1,500 miles per hour, they're only in the air for perhaps ninety seconds at most. No human being could possibly be shaken awake at all hours and asked to validate the system's decision in time to intercept this or that rocket in less than a minute and a half. Too many would get through.

Israel has come to depend on this system, and for obvious reasons. It truly does something better than humans could ever do on their own. But in the process, Israel has also changed the role of humans in conflict. There is now an implicit assumption that in at least one area of military operations, humans aren't just not required, they're in the way.

That assumption is metastasizing across military operations, and around the world. Cyberwarfare, for instance, reduces the allowed response time in combat still further, from seconds to milliseconds, requiring a total delegation of authority to automated systems that detect and respond to attacks in cyberspace. In fact, an entire industry sells automation so pervasive the product categories have their own acronyms, such as SOAR, for security orchestration, automation, and response. As Peter W. Singer, coauthor of *Cybersecurity and Cyberwar* and a strategist at the New America Foundation, put it to me, cyberattacks don't have some hacker on the other end, tapping at a keyboard. "It's software writing software weaponry."

And from cyber to drones to every other aspect of human conflict, automation is changing war, and how we feel about it. Whatever the ethical position the military may adopt, its attitude is changing because the world is changing. Consider, Singer pointed out, that already drone pilots may be driving to work each day in anything from a Tesla to a Ford, all of which come with systems that can alert the driver when they get too close to the next car on the highway. Some of them simply rumble the driver's seat and beep at you. Others apply the brakes and steer. "That's how they'll drive to the military base," Singer said, "and then you're saying somehow in this future world, the minute they walk into the military base, we go, 'Oh, but inside the military base, it's 1999. We're going to fly the predator drones the way we did back in 1999. We're not going to use them the same you use all the other technology.'" Automation has created deep ethical questions in every other domain it has touched, "so don't expect the military to be different in terms of the deep questions it raises, but also don't expect it to be some area that you can keep technology from changing."

Right around the time I became acquainted with my predisposition toward puking and passing out in simulators, I worked on a television show about top-secret airplanes, and I wound up standing in front of a solution to the fainting-pilots problem: the X-47B. Built by Northrop Grumman at a military facility known as Plant 42 in Palmdale, California, just over the mountains from Los Angeles, the plane was the first of a generation of pilotless combat aircraft being tested by various branches of the military. It's an alarming thing to be around. All my twentieth-century instincts about aircraft lead me to expect someone sitting inside, so it's disturbing to see a huge black intake port where the front of the cockpit should be, and big blank stretches of aluminum where you'd expect the pilot to be sitting. The thing literally has no face, no one to whom you can wave your surrender. And yet it has successfully taken off from and landed on an aircraft carrier—one of the hardest things a combat pilot is ever asked to do—without a human involved.

Many thoughts went through my mind standing in the burning sun, looking at the future of war. I thought of how hard pilots in Bahrain had trained to be reliable in an aircraft that I am physically incapable of flying into combat. I thought of how this new plane made the difference between myself and the best trained pilot insignificant. And I imagined this thing pockmarked with gunfire, upside down, cutting a burning trail toward the desert floor. No human inside means neither I nor a Bahraini pilot nor anyone else would have to suffer bullets or burns or bailouts. Maybe that's a good thing. But I also knew that this thing, expensive as it was to build, was a disposable machine. And I didn't know it at the time, but that's where this plane had become part of The Loop. It's a terribly convenient system that will make war logistically and morally and politically easier than it has ever been.

We will make military decisions in that new, psychologically easier world that will go on to be analyzed and condensed and presented to us again in even easier, even less psychologically taxing menus by future decision-management systems. Will we eventually lose any clear sense that we're dealing with conflict, death, murder? We have engineered the human frailty out of this sort of automated system. But how are we going to engineer human morality into it?

ONE MORNING IN 2016, I attended roll call at the Newton Division of the Los Angeles Police Department. Newton comprises roughly nine square miles and 150,000 people across four neighborhoods (including a public housing project) just south of downtown Los Angeles. Police work shifts across night and day, meaning there's none of the sleepy early-morning vibe of a workplace coming to life, and even at dawn the halls were full of bluster and activity as one patrol handed their shotguns and radios across the desk, talking loudly about their plans after work, and another waited to collect equipment for the start of their day.

Roll call began with a handful of routine announcements, and then the watch commander listed off a handful of crimes—vehicle thefts, burglaries, assaults—that a system called PredPol had predicted would take place in their jurisdiction that day.

The watch commander finished by asking the officers in the room to "spend time in the PredPol areas," pointing to a screen behind him, "from the 60s and 70s up and down Broadway." Seated at the back of the room, wearing a Beretta and a tactical vest, Officer Marcela Garcia glanced through the printed map in front of her, which recommended specific times to roll through specific areas. Then she checked out a shotgun and a radio at an equipment desk and walked me out to her Ford black and white, and we settled in together for a drive around the neighborhood.

PredPol is a for-profit company founded by Jeff Brantingham, an archaeological anthropology professor at UCLA who created a system with Department of Defense funding for predicting battlefield casualties on the ground in Iraq. He adapted those algorithms to make similar predictions about crime, using COMPSTAT data from LAPD in collaboration with Deputy Chief Sean Malinowski.

COMPSTAT is a crime-data system created by the NYPD in 1995, since replicated all over the world. It was once praised for standardizing certain forms of data for refining police procedures, but in an anonymous survey of more than 1,700 retired police officials, criminologists found rampant manipulation of the data. "Results indicate that the misuse of the performance management system and pressures on officers from management are key explanations for manipulating crime reports."[1] In one 2010 report, an NYPD whistleblower charged that officers were downgrading felonies to misdemeanors, and in some cases actively discouraging victims from reporting their crimes, to make their COMPSTAT data look better. It was one of the first large-scale examples of The Loop, in my view, when the process of analyzing professional behavior mutated the behavior being analyzed, and not for the better.

In Los Angeles, Brantingham and a research team went on to refine their forecasting process, patented what they built, and founded PredPol, LLC. In 2011, LAPD became one of the company's first signed clients.

More than 450 UCLA faculty and students signed a letter in October 2019 that argued

> to the extent that Professor Brantingham and his UCLA colleagues represent an academic stamp of approval of predictive policing and other algorithm- and location-based methods for [LAPD] Chief Moore, we wish to make it very clear: there is not universal agreement or acceptance of the empirical merit and the ethics of this research.

The letter also raised concerns about the possibility that PredPol's "use of historical crime data naturalizes policies and practices that have had disparate impacts on Black and Brown communities."

By some metrics, PredPol seemed to work. The LAPD Foothills Division reported a 13 percent drop in crime in the first four months after rolling out PredPol. The Alhambra Police Department reported a 32 percent reduction in burglaries and a 20 percent drop in vehicle thefts after it rolled out PredPol. More than sixty departments nationwide adopted the technology, including big cities like Seattle and Atlanta. Atlanta, which adopted it in 2013, reported an aggregate drop of more than 8.5 percent in total crime in the two neighborhoods where it was deployed.

I sat with Professor Brantingham and a camera crew for an interview and asked him whether the system he'd built might fall too heavily on innocent people who happen to live where the math predicts crime will occur.

"There's a big difference between trying to predict *who* is going to commit a crime, and *where and when* a crime is most likely to occur," he told me.

I pointed out that a pattern of where and when can very easily lead to the effect of pointing police at individuals. "I think about a teenage kid who happens to be living in one of the three houses your system has identified," I said to him. "He's going to be more likely to come under the eye of a police officer or, worse, get put into a squad car, even if he hasn't done anything wrong, just because the system has identified that place as a likely place for crime."

He disagreed. "The only data that is being used here is what type of crime is it, where did it occur, and when did it occur. So it's not focused on the individual, it's focused on the events themselves."

"Although it does zoom in on them."

"Not on people, on places."

"On where they live."

"That's right—on where they live."

I spent the morning riding along with Officer Garcia. We settled into the deeply worn black seats of her squad car, and she drove me through one area, a half-block at the center of a five-hundred-foot by five-hundred-foot square on a map, that the software had identified as being likely to see crime at that time on that day. Officer Garcia told me that PredPol more or less tells her when to relax and when to be on alert.

"If we see suspicious activity in the sense of a traffic violation in the area, or a pedestrian violation, we can always do a consensual stop," she says. "Stop and speak to them, get information, of who they are, if they live in the area, and what they're doing in the area."

"Do you guys find that you make more arrests within those boxes? Is that one of the outcomes here?" I asked her.

"That is one of the outcomes," she says. "We saw a reduction in crime just based on the amount of time we spent in those boxes."

I asked Sergeant Garcia effectively the same question I posed to Brantingham: "Does it ever feel to you like for someone who isn't breaking the law but happens to live on this block that we're cruising right now, that they're unfairly getting your eye on them? You're ready for trouble on this block, they happen to live here, but they're doing nothing wrong?"

"We don't want people to feel that way, at least that's not our hope for what they're gaining out of us. We want them to actually feel good that we're more visible in the area in which they live." It may be that some residents of that particular block *will* feel safer by having police come through more often. But for many, especially after the deaths of Oscar Grant, George Floyd, Breonna Taylor, and too many others, that's unlikely to be the case. The objective function doesn't benefit everyone.

Professor Brantingham says the point of PredPol isn't more arrests, just disrupting what the algorithm predicts will happen by exhibiting police presence. "It's been known for quite a long time that there's something called residual deterrence, a police officer shows up in a location, and when they leave, their effect persists for quite a long time." The scholarship for the most part confirms this—the threat of longer prison sentences seems to have no effect on crime, but the threat of apprehension, in the form of more police and more visibility for them, does seem to have a deterrent effect.[2]

But what are the long-term effects on policing itself? That part we haven't studied yet. Like coParenter, the system is doing the training, and as we've seen with cases like David Dao, we're not good at saying no to the system. Imagine a police officer who comes up through the ranks using a system like this. What effect will it have on her patrol habits, her instincts, the broad discretion most departments give their officers, if she's looking to a piece of software to tell her when to turn her vigilance on and off?

After PredPol pioneered the field, companies like Palantir, HunchLabs, and IBM followed with their own predictive-policing software packages, all of which claim, like Brantingham did, that race is not taken into account. But NYU law professor Barry Friedman, writing in the *New York Times* in 2018, put it as follows:

> Police may "go where the crime is," but because so much focus has been on low-level offenses in disadvantaged areas that are ignored elsewhere, these algorithms make it inevitable that the police will return to these places time and again.[3]

To test the company's claims of bias-free policing, machine-learning researchers Kristian Lum and William Isaac ran PredPol's algorithm on a year's worth of drug-crime data in Oakland. Lum and Isaac recorded every time the system, which is based entirely on prior reports of crime, flagged a particular grid of the city as a candidate for targeted policing.

Their results were alarming. Even though drug use is evenly distributed across races, Black citizens were twice as likely as white citizens to be targeted by PredPol, based on the areas it flagged. "We find that rather than correcting for the apparent biases in the police data, the model reinforces these biases." And when they simulated the effects of using PredPol in Oakland, by assuming that police would act on the recommendations of the algorithm and be more likely to stop people, search them, and make arrests, they immediately spotted The Loop. "In each location where targeted policing is sent, we increase the number of crimes observed by 20%. These additional simulated crimes then become part of the data set that is fed into PredPol on subsequent days and are factored into future forecasts," they write. "We find that this process causes the PredPol algorithm to become increasingly confident that most of the crime is contained in the targeted bins. This illustrates the feedback loop."

But here is the real concern, according to Lum and Isaac:

> Whereas before, a police chief could reasonably be expected to justify policing decisions, using a computer to allocate police attention shifts accountability from departmental decision-makers to black-box machinery that purports to be scientific, evidence-based and race-neutral. Although predictive policing is simply reproducing and magnifying the same biases the police have historically held, filtering this decision-making process through sophisticated software that few people understand lends unwarranted legitimacy to biased policing strategies.[4]

In 2019, an internal audit of LAPD's various data programs found trouble throughout. One program, which pointed out "chronic offenders" to officers, was identifying people with one or no convictions about 44 percent of the time. The same audit found discrepancies in PredPol data. And earlier that year, LAPD's inspector general complained he had no way of determining whether PredPol was, in fact, reducing crime, a theme the *Los Angeles Times* discovered in several police departments across the country that had canceled their own PredPol subscriptions. Then, in April 2020, LAPD Chief Moore announced that, faced with budget constraints and the coronavirus, he wouldn't be renewing the department's contract. It turns out that while facial recognition, license-plate readers, and other forms of broad surveillance are on the rise, in the case of COMPSTAT and PredPol, there is some organized, institutional resistance on the part of law enforcement. It may be that agencies only rejected the technology when it began to shape their choices in ways they didn't like. But I take some comfort from knowing that in this case, police across the nation experienced The Loop—fed their choices into a data-analysis system, saw the seemingly useful recommendations that came out the other end, found those recommendations wanting—and rejected it.

In fact, wielded properly, machine learning doesn't have to reinforce The Loop. It might also be able to break it. If we let it, a pattern-recognition system can pick up on the many destructive patterns we ourselves don't recognize and correct them for us. This is what Kahneman, Sunstein, and Sibony argue in their 2021 book, *Noise*. And in theory, that's true.

While algorithms have been relentlessly sold to *serve* law enforcement, a new wave of parallel research is being slowly *applied to* law enforcement. Andrew Papachristos, of Northwestern University, had already found that one's chances of being shot go up dramatically when one knows other people who have been shot. It might seem obvious, but this sort of network science reveals just how small and dangerous the social networks are that contain *anyone* likely to be hit by a bullet. Papachristos's 2015 study of a six-year period in Chicago showed that 70 percent of nonfatal gunshot victims were part of social networks accounting for less than 6 percent of the city's population.[5]

Rather than offer that information as some sort of method of targeting certain citizens for more intense policing, Papachristos and his team instead turned the same analysis on police. A 2020 study of the instances when Chicago police had used force (and received complaints) between 2000 and 2016 found that while only a very small fraction of officers actually fired a gun at somebody, there were clear patterns in who pulled the trigger.[6] Most studies in this vein have looked at either the individual characteristics of police who most often fire their sidearm, or the very local networks (one-to-one relationships with others in the precinct, like the partner or the supervising officer). Papachristos and his collaborators instead used network analysis to sort out the broad professional network of police in Chicago. Think of it as search results on LinkedIn: the second- and third-degree connections that run through the entire city's police force.

What the researchers found is that officers who had shot someone often acted as a social "broker" in the broader network. These were the officers who represented the shortest and most populous social path between two other officers. Often these shooters were officers who had been transferred (or shuffled) between precincts, which likely meant they had social connections in both places. Other factors that seemed correlated to being a shooter included "younger age, being male, getting a larger raise, and having faced many civilian complaints," the study found. Also, shooters were more often officers who'd received complaints along with other officers (meaning they weren't lone wolves; they were connected to other officers and made poor choices *with* them). Race and gender seemed to have no correlation with one's chances of being a shooter. It's not a perfect science, and I recognize it may subject police to the kind of oversimplified data-driven behavior-shaping I wrote this book to criticize, but perhaps it's in life-and-death domains, ones where profit isn't the central consideration, that it's appropriate to consider shaping behavior this way.

But AI-driven reform like this isn't likely to happen across our lives the way AI-driven marketing, political messaging, and curation is. The reason is simple: you don't make easy money on reform. There is no lucrative market for this kind of pattern recognition the way there is for one that helps you find a sales lead or identify a job candidate. But there is a place that capitalist considerations *could* create an incentive. Perhaps the financial strain of lawsuits in the case of police shootings can help to create a market. It could be that pattern-recognition systems could be used to avoid *losing* money.

The family of Louisville medical worker Breonna Taylor reached a $12 million settlement with the city—the largest payout in Louisville's history—after its officers shot and killed her in her apartment. Cleveland paid the family of Tamir Rice $6 million after a white officer killed the Black twelve-year-old. And these settlements not only directly damage a city's finances, they can bleed a city on an ongoing basis, as municipal insurance companies charge more to account for the risk. After a series of police shootings cost the city of Madison, Wisconsin, more than $13 million in settlement costs, the city's municipal insurer raised premiums and deductibles by more than 42 percent, a city manager told the *Wisconsin State Journal*.[7] If those insurance agencies begin to understand that a system like Papachristos's could be used to find the patterns in police training and officer assignments that lead to shootings, perhaps a new market for that sort of AI could result. Because we know that of all the factors that make The Loop go, capitalism is the big one. Perhaps legal restrictions offer the most obvious solution for reining in pattern recognition technology. However, if we can actually invent a money-making way to deploy AI in the service of both justice *and* money, we have a chance to take control of where it goes, and how it affects us.

Chapter 11

WEAK PERFECTION

HOW DO WE put some boundaries around the use of AI in the domains where its use really can determine the course of our lives? First, we must recognize the key assumption we generally make about the wildly transformative potential of artificial intelligence, because it's a dangerous one: that it can improve anything on which we sic it. This has been the moral argument for putting AI into all our lives. The ability to pick patterns out of a vast jumble of data is, of course, hugely powerful. And the resulting predictive analysis allows companies to offer the possibility of getting in front of human behavior, of winning investments, of possible threats. But we haven't yet sorted through the possibility that while robots can turn faster in a fighter jet than we can, that doesn't mean we should let them pull the trigger. Similarly, in other critical areas of life, AI might work against our values, making life more efficient in places where inefficiency is, in fact, a hidden human safeguard.

California Supreme Court justice Mariano-Florentino Cuéllar, writing in an extremely prescient 2016 article, pointed out that if we're going to establish goals for machine learning's role in administrating our lives, "those goals must also inform whether we assign value to the characteristics of human cognition that contrast with what a computer program could accomplish."[1] Computers are good at optimizing a system, but it takes humans to sort out what we actually *want* from that system.

For example, the system of law could of course be made vastly more efficient. "Think about entering a plea," Cuéllar, now head of the Carnegie Endowment for International Peace, told me. "It's a choice you only get to make once—there's no taking it back—and that plea has the potential to change your life forever." Entering a plea in California is a rigidly regulated process. The California penal code stipulates that pleas must be entered in open court, for one thing—no doing it remotely. And if I'm going to plead guilty or no contest, the court has to provide me additional time to consider the implications. And if the offense carries the risk of deportation, a life sentence, or other extraordinary punishments, all sorts of warnings and negotiation processes and approvals from my lawyer become necessary as well.[2] Pleading guilty or not guilty is extremely complicated.

Could technology make it simpler? Of course. We might reduce the whole process to a single swipe on a smartphone, a world in which defendants conveniently and casually enter their plea one-handed from a cell, or while waiting at a stoplight. But that would be disastrous, according to Cuéllar. "The legal process puts people through an intentionally slow decision-making system when they decide on their plea. It forces slow thinking. We call it 'weak perfection,' the idea that the system remains difficult to help them make better decisions."

When I think about how I'd want pattern-recognition algorithms to be regulated, I think about how "weak perfection" describes a whole host of systems that are slow, inefficient, fragile, difficult to navigate—and force us to stop and do the most mentally taxing kind of System 2 processing. Perhaps sometimes we *should* have to do it the hard way. We *should* have to write out our specific desire not to be intubated at the hospital if we fall into a coma. We *should* have to sit and decide how we want our money to be distributed among our children when we die.

Meanwhile, AI is making its way into certain systems of judgment to make them far more convenient for the bureaucracies they serve, and in exchange we're losing our insight into how important decisions are being made. AI can find patterns and form predictive rules, but as we've seen, the decision-making mechanism is hidden from view. Professor Michele Gilman, who teaches law at the University of Baltimore, has spent two decades suing on behalf of people living in poverty who have been denied their rights and benefits. And increasingly, she's found that algorithms are at the center of her cases. In a 2020 paper for *Data & Society*, she offered a guide to the algorithms now being deployed to make decisions in literally dozens of areas of American life, from credit scores to public benefits to housing. Gilman points out that algorithms often make life-changing decisions without directly consulting the people whose lives they're changing. "An applicant for rental housing, for instance, may be denied based on an algorithmic determination in a tenant screening report," she writes, "yet never learn about the basis of the denial."[3]

Gilman identifies several increasingly alarming layers of opacity. Except in a few progressive jurisdictions, like Seattle and New York City, which require transparency and public consultation before an algorithm can be adopted, government agencies usually activate these decision-making systems with little or no public input. And she points out that although due process usually requires that the companies making algorithmic products have to explain them in court, that tends to involve a fight because those companies usually force government clients to sign nondisclosure agreements and claim trade-secret protection when asked about the inner workings of their AI.

In fact, just standing in court, charged with a crime, you or I would likely be under the gaze of an algorithm. Pretrial risk assessment tools (RATs) are in use in forty-six states and the District of Columbia, according to a report by the nonprofits Media Mobilizing Project and MediaJustice. These tools were introduced as a means of fighting off the inherent inequality of cash bail, which disproportionately keeps poor Americans in jail, and as a means of compensating for the possible biases in a judge's discretion. (One influential National Bureau of Economic Research study found not only that Black Americans were disproportionately stopped, searched, arrested, and sentenced harshly, but that those racial biases were consistent in both Black and white judges.[4])

RATs look across variables in a population to offer a statistical prediction of the likelihood that a defendant is a flight risk or might commit another crime while free awaiting trial. Those variables might include demographic information like age, access to housing, and job history, and might also include criminal history, such as substance abuse, prior failures to appear in court, and time spent in prison. The software produces a risk score for each defendant so that judges have a handy way of evaluating whether to release them pending trial, ostensibly free of their own biases.

But as several investigations and research projects have found, RATs can be deeply biased themselves, whether because they channel the mathematical inequalities inherent in American life into their judgments or because they create an overly broad judgment from millions of cases that may not apply to a local court and the conditions facing defendants there. Even studies that have found minimal racial biases in RATs come with stark warnings about overreliance on the software. "Risk assessments are believed to remove conscious and unconscious forms of human bias, and provide a system to treat people fairly," wrote the authors of one *Criminology & Public Policy* study that found RATs don't themselves wind up having an unfair impact on defendants. "However, there is nothing inherent in risk assessments that will reduce jail populations, make prison populations less racially disparate, or otherwise reform the criminal justice system."[5]

Some who hoped RATs might unwind bias in the criminal justice system have now come to actively argue they shouldn't be used at all. Back in 2014, when the state of New Jersey was considering whether to adopt RATs, an endorsement from the nonprofit Pretrial Justice Institute helped to persuade lawmakers to use them. But in 2020, PJI, whose mission is to improve the equity of the criminal justice system in how it evaluates defendants, wrote in a statement that it had been wrong. "We heard but did not fully appreciate the opposition to pretrial risk assessment tools from civil rights organizations, impacted people, and researchers," the statement read. "Regardless of their science, brand, or age, these tools are derived from data reflecting structural racism and institutional inequity that impact our court and law enforcement policies and practices. Use of that data then deepens the inequity."[6]

The tendency to broadly apply the rulings of artificial intelligence when human systems are overwhelmed by human need damages independent judgment everywhere—not just in the courts. And it's when we strap AI and high expectations onto a broken system that things really go wrong.

By January 2021, ten months into the COVID-19 pandemic, the economic fallout in the United States was severe and deepening. The service industry was decimated as restaurants, hair salons, and gyms faltered and closed. The head of the National Association of Consumer Bankruptcy Attorneys told me his members were preparing for an avalanche of lost homes and cars as American entrepreneurs everywhere threw whatever they could at saving their businesses. And as the New Year began, with more than 10 million jobs gone from the economy, more than 9 million Americans fell into poverty.

At that point, a stimulus bill was supposed to begin sending checks to help Americans in need. But because the administration of that relief had been handed off to the states by the Trump administration, state unemployment offices were in charge of distributing those funds. In California, that agency is the Employment Development Department, and like all such agencies across the country, it was utterly overwhelmed.

In 2010, the worst year of the prior recession, EDD saw 3.8 million unemployment claims. But the coronavirus brought more than *16 million* claims, and as of October, 600,000 calls per month meant waiting on hold for an entire working day in the hope of getting through. And then, in November, news broke that EDD had been accidentally sending benefit checks to people in prison. It turned out that since the beginning of the pandemic, nearly 40,000 inmates had filed claims, federal and state prosecutors found, including 133 death-row inmates, and more than 20,000 such claims were paid. All told, the state had wrongly given out more than $140 million to incarcerated people. (Some might argue that incarcerated people need additional resources, but surely distributing millions by mistake only to those who thought to illegally apply for them isn't the way to do it.)

The EDD, reeling and embarrassed, and without the political support to ask for enough humans to clear the backlog and process new claims, decided to use a risk-detection system.

The first sign that this new system was going to pose problems for everyone was a tweet by EDD on a Sunday in January 2021: "As part of ongoing efforts to fight fraud, EDD has suspended payment on claims considered high risk and is informing those affected that their identity will need to be verified starting this week before payments can resume. More details on the EDD website in the days ahead."[7]

I asked EDD how they were determining high-risk claims. They told me they did it by throwing an off-the-shelf piece of software at the problem. "Using the Thomson Reuters fraud criteria, EDD reviewed existing claims and applied additional industry standard fraud detection criteria, taking action on claims deemed highly suspect or fraudulent," they wrote. "In order to mitigate future fraudulent payments, EDD has stopped payment on those higher risk claims and has communicated to those affected that we will be requiring identity verification or additional eligibility determinations starting later this week before payments can resume."

How many claims did the Thomson Reuters product flag? *One in seven.* More than 1.4 million claims were frozen by the system, meaning those Californians, at a time of extraordinary hardship, would have to navigate additional bureaucracy to prove they were eligible to collect a weekly payment of $300 from the federal government. The process reeks of misplaced assumptions. EDD assumed that software could better decide on a course of action than could humans. The software assumed that one in seven Californians were in some way undeserving of immediate aid. And EDD assumed the software was right. As we hand more and more crucial decisions to automated systems, the assumption that computers can do things better than humans can is going to become a matter of life and death.

Perhaps we've simply gotten ahead of ourselves. Perhaps inside The Loop we're just too quick to measure the benefits of AI in time and money—and avoid uncomfortable choices in the process. Perhaps where AI should be applied is not in issuing predictive judgments, which creates deep inequities that are difficult to untangle once we've built a human system on top of them, the way EDD did.

Instead, the same way we've considered using AI as a potential way of looking at what's wrong in policing, what if we use AI to assess our ugliest hidden patterns in all parts of life?

In 2018, I toured the Metropolitan Museum in New York with its chief digital officer at the time, Loic Tallon. The museum was in the process of digitally scanning its collection as part of a vast initiative to make it easier to experience both in person and online. For example, as part of a partnership with Microsoft, the Met was beginning to make recommendations to visitors as to what they might enjoy next, the way recommendation algorithms identify the next item you're likely to buy on Amazon. Tallon and I wandered through the Byzantine section and riffed on what AI might be able to do with the collection as a detailed data set. "We've tagged 5,000 years of human history," he told me proudly. And while making recommendations to visitors is cool, he said he was excited about the possibility that the museum could use machine learning to "genuinely *study* the collection itself, and start finding trends and pointing to particular things that the AI may spot across multiple centuries of content." For instance, he theorized, the AI might be able to fill in certain evolutionary gaps in the history of art. Pottery, he pointed out, is so inherently fragile that there are huge periods of time in which we simply don't have representative samples. In the time between Egypt's fall and the rise of the Greek state, one jug and the next known example could be more than a century apart. But AI could look at the materials, the painting, the proportion, and speculate as to what might have come between the two. "It feels inevitable that at some point the AIs will be sufficiently trained, the data sets will be sufficiently large and rigorous, that genuinely you're going to find those patterns," Tallon said. We won't be using AI to sell art, or to make it. We'll be using it to better understand our own creative impulses, what connections our society tends to make between aesthetic and artistic movements. We'll uncover the hidden topography of human history.

Whole worlds of scholarship have been waiting for this sort of tool, a way to rigorously identify patterns in our past as clearly as AI identifies cancer in millions of MRI scans. But can we somehow give it to the people who can make ready use of that tool in studying human beings, without also giving it to companies that want to use those same studies to identify markets and sell products?

AT THE END of a long, isolated gravel road in the woods of Virginia, there's a quiet couple that studies the very worst human behavior. The historian Roberta Senechal de la Roche is a professor at Washington and Lee University in Lexington. She's a poet and has written a book about Civil War snipers (an ancestor was a Union sharpshooter), but her life's work is the analysis of collective violence. She specifically studies the group decisions that have led to lynching, race riots, and terrorism, and the conditions that made lynchings so common and race riots so unusual in the South, while the opposite was true in the North.

Her husband is the University of Virginia sociologist Donald Black, who studies what he calls "the behavior of law"—how it reflects ancient tendencies and why it falls so heavily on the least-powerful members of society. I have been fascinated with the two of them for years, because to me they embody two things that are both contradictory and complementary. On the one hand, they seek to study long-term patterns in human behavior, because as a society we've been terrible at seeing those patterns clearly. They know better than anyone the myopia that will keep us from spotting technology's slow, steady influence on how we live until it's too late. On the other hand, their work is the sort of scholarship that depends on more data than humans can possibly process. If these two were given pattern-recognition systems, and applied them to the history of lynchings, or the history of law, histories that involve more massive amounts of data, the work could transform our understanding of ourselves. But people are using AI to recommend eyeglass frames and vintages of wine. No one's hoping to make billions of dollars by paying to train AI on the history of violence.

Black's work in particular could benefit immensely. "I've always been interested in the contrast between the tribal way of life and the modern way of life," Black told me. "These clashes of 'right' and 'wrong'"—the crimes we have committed and punishments we have meted out throughout history—"how do they change between the tribal and the modern world?"

Black is not a modest person. His theories are vast and sweeping. An academic journal once asked to interview him, and because, as he told me, no one else was qualified to interrogate his ideas, he submitted a *self*-interview. And his work, which he says draws on evidence from all over history and all over the world in service of theories that apply to all of humankind, was called by one frustrated critic, the NYU sociologist David Greenberg, "a terribly inefficient way to generate the elements of a theory."[8]

But there is something haunting in what Black has spent his career describing, and if AI were brought to bear on it, it might be extremely valuable. His theory is roughly this: the way we assess crime and dole out appropriate punishment today is a translation of ancient moral codes. Inside those codes are some very fundamental assumptions and biases when it comes to power, gender, and tribal identity. And those codes play out over and over again in how societies deem certain behavior criminal and accord it more or less punishment. The overall rule of thumb has to do with what he calls "relational distance"—how close or far from the center of a society's tribal and moral identity are the perpetrator and the victim. The severity of the punishment increases as the perpetrator becomes more distant, and as the victim becomes closer to the center of the society and its power structure. At first what he describes seems head-smackingly simple. The killing of animals is punished much less severely than the killing of humans, for instance. But as he proceeded along that path, he found that the consistency with which some crimes are punished more or less harshly reveals how societies rank the value of the people committing crimes, and the people suffering them.

"I don't know of a single case of a man killing his wife in which the man received capital punishment," Black told me. "I mean, there must be some case somewhere in the history of capital punishment in the United States, but I've never seen one. You have [capital] cases where a man kills, say two or three members of his family, but not just one, not just a spouse." The code by which we punish men who kill their wives tells us something horrible, Black told me, about how we feel about marriage, and the privileges of men, and what dominion they get to have over their wives. Black says it may also suggest we prize civil relationships between strangers, and perhaps the lives of children, over the life of a married woman. "Killing an intimate attracts drastically less severity than killing a stranger," Black said. "Capital punishment is basically a punishment for stranger killings, plus multiple killings."

And law, according to Black, has its own gravitational rules. Pour it from a higher sociological height—from a more highly placed victim in society down toward the perpetrator—and it falls more heavily. People who commit a crime against the state are far more severely punished than when the state commits a crime against citizens. Black men convicted of federal crimes receive 20 percent longer sentences than white men convicted of the same crime, according to a report from the United States Sentencing Commission, and that disparity is only growing.[9] And the race of the victim is hugely consequential. A Northeastern University School of Law review concluded that "regardless of the perpetrator's race, those who kill whites are more likely to face capital charges, receive a death sentence, and die by execution than those who murder blacks."

When I asked around among law professors and sociologists, Black has the reputation of being on to something without having produced the evidence necessary to support it. "It's the kind of thing where he might be right, but he can't prove it," one sociologist told me.

Black says he's frustrated in his work, because he feels he's nailed a really compelling theory, but not enough people are paying attention, and he doesn't have the tools he needs to flesh it out. "People say 'but why is that? Why does more relational distance attract more law?'" he said. "I can tell you why the cases are handled the way they are with this theory. I can't tell you why the social geometry of the cases predicts the way the cases will be handled. Maybe someday someone will figure out how." He sighs. "I feel like if some lab professors could just understand it, they could make a whole damn career off of it."

What Black has inadvertently done is build a theory the same way AI might do it—he's taken a pattern-based shortcut to an answer and has built a broad theory on it. But now he's facing the explainability problem. He doesn't have what it would take to go inside and figure out why his theory is seemingly so correct, or what else it might predict.

In a court of law, the most powerful use of AI might be as a mirror, one that shows us the mathematical patterns of inequality we've perpetuated across hundreds of years and millions of cases. And if a hypothesis like "relational distance" is borne out, perhaps *that* becomes the basis of a recommendation to a judge, rather than the sort of systems that now dress the brutal mathematical inequalities of modern life as impartial, time-saving judgments, and pour them down onto defendants.

That sort of problem might be just the thing for AI to accomplish. Feed an algorithm enough of the history of criminal justice in the United States, and it might very well begin spitting out certain predictive connections. Perhaps a theory like Black's can be investigated the way AI might identify the missing links in the history of art. If we're going to blindly believe what AI has to say about something, maybe it should be about who *hasn't* received justice, rather than who should. Perhaps that *needs* to happen before we put AI to use in determining whether someone gets out of jail tonight.

Chapter 12

HIGHER MATH

UNTIL NOW THIS book has offered a few theoretical solutions to all the problems The Loop presents. They're invisible problems, years out, perfectly tailored to our specific human biases and heuristics, and to the business incentives of capitalism. It's a dark picture, I know—I've been ruining dinner parties with it for years. But I want to end the book with some possible solutions. Here we'll look at a few projects that harness the same technology and business dynamics that make The Loop possible, and could conceivably be the beginning of a Loop from which we might actually benefit. "Harness" is the term because until now we've allowed these forces to simply run wild. But with some work—a hard look at how the brain does what it does and at the things it doesn't do well, as well as a similar inspection of AI and of capitalism—I think we can build some rules around this thing and break the circular pattern of guidance-for-profit. It looks impossible, but there are shoots of green already appearing across the landscape.

After years of record-breaking wildfires across Northern California, major insurers began to drop their coverage of certain homes outright. By the math of their risk-assessment systems, the gamble simply wasn't worth it. A company called Verisk, which dominates the data-analytics market for insurance underwriting, estimated in 2020 that 15 percent of properties in California were at high or very high risk for fire.[1]

The traditional means of assessing fire risk is a low-resolution method: average risk across a broad number of homes, whether by evaluating census tracts or entire zip codes. And by that system, there's little or no room for negotiation, because an insurance company can't afford to send someone out to every home multiple times to check out what fire-proofing improvements have taken place. All of this meant that a homeowner who had cleared away brush and thinned trees and put fire-safe vents around her home was being lumped in with her neighbor who still had dense, flammable brush everywhere and siding and roofing that made his home likely to catch when the next embers floated down from a forest fire nearby. Insurers had no choice but to lump the two together, based on historical patterns of fire risk, and deny them both.

But Met Life, Farmers, and a few other companies have begun quietly offering *more* insurance in Northern California communities that are at risk of wildfire. How? They've bought technology from a company called Zesty.ai, which uses machine learning to find the specific factors that endanger a home in a wildfire-prone area.

Attila Toth, Zesty.ai's CEO, says the current risk-assessment models used by insurance companies are outdated. "Of the five largest loss years, four are in this decade, but the technology to understand those sorts of disasters is thirty to forty years old."

Toth's company feeds satellite imagery and property databases through machine learning and builds a recipe for predicting loss.

"We pick up thirty property-level modifiers, things like roof material," he says. "We measure overhanging vegetation from an aerial image. We look at the last building permit pulled for that roof. Slope, vegetation, precipitation, orientation of the property." Then, with that recipe, his AI builds a predictive model that gives Met Life an automated system for offering specific assessments on a property by property basis. Remember that UC Berkeley study, which theorized that one of the ways to fight "performativity" in AI was to stop making all predictions on a backward-looking basis? Instead, the authors wrote, AI should evaluate its predictions "against the future outcomes that manifest from acting on the prediction." That's what's happening with these homeowners.

The result is that insurance, rather than being a binary up-down vote on you and your house's chances of disaster, could become a means of spurring humans to take better precautions. Your trees are too close to the house, your insurer might say. There's dense brush inside the property line, the flashing around your roof isn't fire-resistant, and so here, here's your estimate. But if you bring a chainsaw and a contractor to bear on your fire risk, here's a vastly more affordable quote.

Toth says that his AI-driven model estimates only 2.5 percent of properties are at risk, as compared with Verisk's estimate of 15 percent, and while a narrow slice of that group may have to be prodded by lack of insurance into living elsewhere, a lot of them may be able to take the steps his AI and his inspectors have identified to lessen the danger. "Even if a property falls within the fire perimeter," Toth says, "if you have defensible space, if it's well-maintained, if it's oriented in the right direction, the probability of that house burning is very low."

Of course, this doesn't solve the larger problems of risk and analysis and profitable operations that plague the insurance industry. It's not as simple as saying that we need to use more humans in the process, for instance—that's just not feasible. And generally speaking, an industry built on using math to aggregate human behavior into pools of risk, and then betting on the outcomes, is not looking for exceptions, it's looking for rules. It doesn't want to tailor a suit for each customer; it wants to sell one suit. And so far, none of the companies using Zesty.ai are issuing tailored suggestions to their clients—you have to know to push for it when they issue you a quote or turn you down. But in a world in which the climate is changing, fires are becoming more frequent and more destructive, and people are doing their damnedest to adapt, an insurance company that only uses historical data and lumps everyone into the largest possible pools is signing up for risks no balance sheet can handle.

It's not at all clear that whatever companies do with something like Zesty.ai's system will make life easier for everyone. The arc of history doesn't bend in a satisfactory direction for all customers. Many of the people who live in the path of wildfires don't have the resources to bring in a timber team to trim back their trees. Further, an automated, tailored set of "or else" recommendations may force them to either live without insurance and risk losing everything when fire comes through, or give up the house ahead of time. And how do you sell a home that cutting-edge technology has determined is currently indefensible?

But this is a case where some forward-looking, AI-powered analysis can help us perhaps do what we've never been able to do before: encourage people to think about their risks before they buy or build a home. There is currently so little regulation as to where one can build a rural or semirural home in the United States, and there is so much climate-fueled catastrophe coming toward us, perhaps an AI-driven system for determining risk is a necessary piece of guidance. In California, homes are regularly built in the path of recurring wildfires, in the shadow of levees across the Sacramento–San Joaquin Delta, or in low-lying oceanfront communities in Los Angeles, all of which are subject to the intensifying rhythm of fire and flood and coastal storms. Zesty.ai could apply its technology to any disaster-prone location in the world. And if it did, perhaps we'd wind up pulling people away from places that simply aren't safe to live any longer. AI can help *sell* a home by matching historical patterns, reading emotions, even writing advertising copy, sure. Those are the forces of The Loop. But if AI can help *dissuade* us from buying home in a dangerous place, or ensure that we don't sign a mortgage on a home without understanding that we'll need to spend another $10,000 fireproofing it, that's something else.

IN 2009, I attended a wedding in Antigua. Antigua is a cobblestoned former colonial capital in the south of Guatemala, surrounded by volcanoes. My wife, who is more adventurous than I am, convinced me we should hire a guide to take us to the top of one of them.

The hotel staff told us the best way to see a volcano in the area was to schedule it such that we hit the summit at sunset, so the lava would be glowing and visible. "Lava?" I remember thinking. The next afternoon we met a fit young man in hiking boots outside our hotel, where he drove the two of us and a half-dozen other tourists to the base of Volcan Pacaya, a nearly nine-thousand-foot peak.

I am naturally vulnerable to the effects of high altitude—I sleep poorly, I become afraid of the dark, I can't keep thoughts straight in my head—and my natural anxious vigilance was already pumping. I glanced at my own sneakers, wondering if they'd hold up, and worried that my sweatshirt and shorts wouldn't be enough to keep me warm up there. My wife would be fine, I knew—she's frighteningly tough and capable in any situation. Then, as we milled around waiting to set out, I took stock of our group. The father and his teenage girls seemed hearty enough, and they were all wearing sneakers. But the newly married couple from Brooklyn seemed winded just standing around, and the wife was wearing brilliant white socks in a pair of one-strap slides.

We set out on what proved to be a fairly easy climb, but as we passed 4,800 feet, the vegetation disappeared, and we were picking our way through a loose clattering of cooled lava rocks. At one point I stumbled, putting my hand out for support, and drew it back bloody. The rocks around me were dangerously sharp. And when the strange, jagged moonscape leveled out near the peak, we were suddenly among rivulets of glowing lava, watching it twist and move across a landscape that was already burning my feet through the soles of my shoes.

There were no rules up there. I saw tourists from other groups leap across rivers of molten rock whose heat made me squint from ten feet away. I watched a young boy walk toward one pool with an extended stick, only to have it burst into flame before it even made contact. And three peaks distant, poking through the clouds below us, I could see a column of smoke from another active volcano rising on the horizon. That's when I realized that all of this was uncertain, that we were essentially atop an unstable archipelago in a sea of fire, and that our footing could give way at any moment and plunge us into unimaginable pain and injury. I'd been all uneasy gallows humor with my wife until then, but at that point I looked her full in the eye, the wind howling around us as darkness intensified, and said, "This is crazy. We cannot stay here."

With some urging from the two of us, our guide began to rally our group to depart, and he asked us if we'd brought flashlights for the way down. This was news to all of us, and I weakly held up my phone, with its dim flash. It seemed that it and the guide's flashlight were the only means of picking our way back through the sharp rock field, on a moonless night, where a fall would mean laceration or worse.

I have often been in the position of describing to executives at various large tech companies my worries about the long-term effects of their work on human behavior across generations, and when the conversation turns to what should be done about it, I often trot out something about the difference between being *able* to make money a certain way and the decision to go ahead and do it. And that's when surreptitious eye-rolling begins, along with talk of scale and shareholder responsibility and the positive and negative uses of any tool, and sometimes true outrage at my suggestion that companies should perhaps be forced to stifle or even dismantle portions of themselves. Soon I am dismissed.

It is one of the founding assumptions of American capitalism that if someone will pay for a service, then some form of that service is a legitimate product, and selling that product is a worthwhile business. Throw in the free market and competition, and capitalism will level itself out, the thinking goes. But there is an invisible layer of protections we have bolted on. In tourism, for instance, we have come to assume in the United States that if someone is charging money to take you somewhere, it's safe to go there with them. But as I descended the sharp slopes that evening, the wind growing colder on my bare legs, the guide trying his best to take the arm of the honeymooner fighting bravely to keep her footing in her slides, her white socks now dirty and frayed, I realized that while this guide and his peers were in the business of taking tourists to the top of the volcano—a business that was undoubtedly feeding his family, perhaps his friends and neighbors, too—we had no business being up there at all.

Why not? Because no infrastructure existed to modulate the relationship between us, tourists willing to pay American dollars to go somewhere exciting, and him, a guide with comparatively few opportunities to earn comparable money. The top of Volcan Pacaya was a terribly dangerous destination, and yet no one was weighing that against the income to be made leading us up there.

In fact, just four months after I returned home from that trip, a Venezuelan tourist and her guide were killed on the same hike we took, when the tenuous surface of that rocky slope burst open, pouring steam and fire and a fusillade of burning rocks over the two of them. And an eruption a month after that showered three towns in the area with what volcanologists call bombs—large, semi-molten chunks that splatter on impact. The eruption killed a television journalist and two others. Photos of the aftermath show the interior of a church with its corrugated-metal roof perforated like a colander with holes a foot across.

I have grown up in the sort of American environment in which railings keep you from walking off edges and fences keep you from getting too close to the lava. But it took this hike, and the realization of what might have happened if my wife and I had gone on a different day, to make me realize that just because someone will take you somewhere doesn't mean it's safe to go, and just because someone is willing to make a purchase doesn't mean that thing should be sold.

In the United States, it has taken more than a century of battles in court and in government to establish building codes and liability law that slaps railings and fences around us as we go about our lives. These systems are not perfect, and there are enormous dangers in the United States—firearms, industrial pollution, alcohol—that public-health experts have argued for generations are woefully underregulated. Still, we sign elaborate waivers that detail the risks to us before agreeing to go past the safety barricades, and even then, negligence or obfuscation still gives an injured person grounds to sue. And here's the thing: on Pacaya it turned out there was some small regulatory safety net at work around us. But it wasn't protecting us from lava. The guide told me later that his worry on our descent was not the possibility of eruption, or that we might fall and cut ourselves. It was that if it got too far past darkness, the guards posted along the trail—bored-looking men with shotguns and rifles who were paid to deter thieves—would go home, and we'd be easy prey. Whatever system loosely oversaw tourists on that volcano didn't seem to be built to minimize the risk that we might be maimed by the volcano. But that system certainly recognized the more likely danger, that we'd be robbed once the sun went down, and paid a half-dozen guards to keep that from happening.

The economist Albert O. Hirschman describes three options available to modern citizens faced with a system that does not serve them in his 1970 book *Exit, Voice, and Loyalty*. Don't like something? Hirschman argued you have three choices: walk away, argue for improvements, or stick with it. Exit, as he describes it, is an economic mechanism, a highly efficient error-correction system for improving the performance of companies and markets. Voice, the messier option, is a political mechanism. Loyalty, in his model, is the sweetness that makes exit more powerfully sour and threatening. And with the concept of exit, he believes it can be tremendously useful if a bit of elasticity is built into the market: a mix of astute and inert customers, as he calls them. The astute customers, noticing the problems with the business and exiting as they do, will provide the pressure toward improvement. But the inert customers, who don't notice what's going on and keep paying, give the business enough stability and resources to make the improvements that the astute customers have asked for with their departure. "It is important that other customers remain unaware of, or unperturbed by, quality decline: if all were . . . determined comparison shoppers, disastrous instability might result and firms would miss out on chances to recover from their occasional lapses," he explains.[2]

A trip to the top of a volcano by customers like me and my wife who don't have the means of researching the danger in any meaningful way, and who will likely never make the trip again, means we couldn't exercise voice until it was too late, and knew to exit ("I'll never do that again") only once we'd paid the fee and returned from the volcano. There are some systems that simply cannot self-correct. The Loop is one of them.

A behavioral-guidance system deployed at scale means that the vast majority of people will be inert, and even the rare astute customers will be corralled by the product, without any hope of making an exit. Corporate customers, the ones buying and deploying the algorithm, don't have the same incentives as the end users (you and me) to inspect anything beyond the immediate financial impacts that AI's efficiencies offer their business, so we can't rely on them to use voice or exit for us. And if an entire category of products is being organized and fed to us by pattern-recognition software, such that the results are just good enough to satisfy us, we'll be saturated by the result, and sedated by the convenience. What possible means or incentive will we have for looking inside how that works, differentiating between one outlet or another, and organizing some sort of customer resistance to it?

The Loop isn't subject to what Hirschman called "recuperative mechanisms" because it creates financial incentives no other system can beat. As such it will wind up becoming a sort of technological monopoly that offers short-term benefits and conveniences no one has the vision to see past. You've searched for a recipe online, I'm sure, which invariably subjects you to a long diatribe about the author's recent trip to Brazil, or an endless childhood parable, before you can get to the instructions you're there for. It's not some sort of irritating, self-important custom among kitchen influencers. It's that Google's algorithm penalizes short-form recipes in its search rankings. In order to get found online, the authors *have* to write a lengthy block of text. You find it annoying, but you still look up recipes online. They're free, after all. And in the end that becomes a normal part of an afternoon at the stove. Recipes buried at the bottom of a breezy journal entry are no longer confined to a handful of blogs you can choose to either email a complaint about or abandon. It's the whole ecosystem.

Hirschman described this as a "public goods" problem, in which people partaking of a common resource—national defense, public schooling—can't functionally "exit" that system without moving to the woods entirely. The Loop has the potential to make all aspects of life subject to AI-driven selection mechanisms, and as a result we won't be able to object or leave. Loyalty may wind up being our only option.

THE LOOP IS going to fool us into believing that AI is worth adopting for the time and labor it saves, and the case for its use will be made again and again in the numbers. A Facebook CIO famously used the phrase "data wins the argument" to describe the deliberative process inside the company. But I believe that we should, as a society, prioritize certain things above expediency, and that AI cannot be trusted to absorb our *values* in any reliable way, because often acting on our values is going to require acting in inefficient and costly ways. Frankly, doing the right thing often feels wrong, which is why there are other ways (courts, regulations, philosophy) to evaluate the decisions we make as a society, and before we go handing much more of our lives to AI, we should make sure we've taken stock of who we wanted to be in the first place.

Sometimes we can express our values as math. In general, America has relied on simple mathematics to determine the value of a human life, for instance. Federal agencies in fact price human lives at a fixed rate—the VSL, or value of a statistical life—as they calculate the financial impact of a regulation. (EPA's rate is $9.1 million per life, whereas the Department of Transportation pegs it at closer to $9.6 million. Strangely, VSL in other countries is lower, presumably because Americans are more litigious, or perhaps because we enjoy such an enormous GDP. A New Zealander killed in a car accident is worth less than half what an American is according to the NZ Transportation Authority.) There are cases where the gravity of our loss is expressed clearly in the math, and the remedy is easy to measure.

But then there are instances where the gravity of loss isn't just a matter of raw numbers, it's a matter of secondary losses, or even of emotional impact. As my friend and producer Carl Byker pointed out while we were filming a documentary together, we have decided, as a society, that certain extraordinary remedies are necessary when a loss is just too much—emotionally and psychologically speaking—for us to bear.

Sometimes it's a matter of building unique *exceptions* to our rules so that our civilization can continue to enjoy the benefits of a thing that in rare instances hurts individuals. Vaccines are an example.

I want to be clear, up front, that I think vaccines are the greatest lifesaving invention in human history. This isn't just because they're saving lives from COVID-19 all over the world as I write this. Whether it's an inoculation against polio or tetanus, only water purification and sewage systems have served to fight off death and debilitation as well as vaccines have. Life without them would be short and end horribly for billions of people, and being vaccinated is the ultimate act of communal compassion because it protects not just your body, but the bodies of your family, neighbors, colleagues. When a vaccine passes clinical trials and makes it to market, get the shot.

That said, there are rare cases in which people are harmed by a routine vaccine, typically because of a severe allergic reaction. How rare is this? The CDC estimates that for most vaccines the rate is literally one in a million, and in most cases the reaction is anaphylaxis, the swelling and breathing obstruction allergic people get from a bee sting. That usually happens within four hours, and that's why the healthcare providers who administer vaccines are equipped with ready shots of epinephrine, which can reverse the allergic reaction almost instantaneously.

But there are cases in which something like thrombocytopenic purpura (a crash of one's blood-platelet count, leading to bruising or uncontrolled bleeding) or Guillain-Barré syndrome (a debilitating form of nerve paralysis) results from vaccines. And this puts our society in a bind. The number of deaths and injuries from routine vaccines is exponentially fewer than the deaths and injuries from the diseases these vaccines prevent in the rest of the population. But the injury to this *extremely* small number of people each year is bad enough (and, as it typically happens to children, emotionally horrible enough) that conventional liability law would make it impossible for any pharmaceutical company to continue taking the risk of producing these vaccines. Still, we need those vaccines. The billions of lives saved each year means we have to keep them in production. And so we've set up a separate legal system entirely, a carved-out exception to run-of-the-mill capitalism, in order to make vaccine production legally viable.

If, as a parent, your child is injured by a vaccine, and you choose to pursue damages, you can apply for remedy at the National Vaccine Injury Compensation Program, which hears cases at the Howard T. Markey National Courts Building, on a leafy street in Washington, DC, that looks out over the grounds of the White House. There, a "special master" will preside over your case, hear your evidence, and, assuming that medical staff from the Department of Health and Human Services find that you fall into the proper category, award you a payment, almost immediately, for your and your child's loss.

The payments cover medical and legal expenses, future wages lost, up to a quarter-million dollars for pain and suffering, and another quarter-million, God forbid, in cases of death. Even those who file an unsuccessful claim usually get their legal expenses covered. It's an extremely efficient and sweeping system. The program has paid out more than $4 billion since it began in 1988. And it's a radical departure from the typical rules of American business and tort law. The vaccine makers (and American citizens) fund the pool of money through a 75-cent tax on every vaccine dose administered. And in exchange, those manufacturers enjoy a "no fault" status in those cases, avoiding legal liability for making vaccines that have harmed someone, so everyone else can live in a vaccinated society.

Not only is all of this a radical departure from how capitalism handles nearly everything else, it reflects a very new way of thinking about what children mean to us. The sociologist Viviana Zelizer notes in her book *Pricing the Priceless Child* that the death of a child was considered until the 1930s a comparatively minor hardship. Today, although a child typically doesn't work a job the way children had to back then, a child's life is considered vastly more valuable, against all economic logic, and compensation for the loss of a child today reflects that illogical yet inarguable society-wide consensus. The point is, we can invent new systems for new values when it's important enough to do so.

These compensation programs that do an end run around capitalism are a model for the sort of thing we'll have to invent for new forms of technology. Back in 2007, when self-driving cars were first becoming technologically viable, I interviewed legal experts on the implications of a vehicle that would so clearly shift legal responsibility for any accidents from the human driver to the manufacturer. Even then, the experts I spoke with were convinced that robots would inevitably be better at driving cars than humans. (One World Health

Organization study estimated that drivers are at fault in more than 90 percent of accidents.)

"If," as a car manufacturer, "you can drop the number of accidents by taking over control, you're going to face fewer lawsuits," Ryan Calo, who was studying the issue at Stanford at the time, told me. He said he thought car makers would be happy to take legal liability because the pool of liability would still be smaller than it is today. "If you reduce accidents by 50 percent, who cares if you hold the bag on the remaining lawsuits? You've still saved yourself a lot of money."

If we get the technology right, and the number of accidents would be so much fewer, perhaps it's worth thinking of autonomous vehicles as an inoculation against human error, a literal vaccine against bad driving. We have a vaccine court for flu shots. Perhaps we'll have to build a vaccine court for robot cars and other forms of AI-based technology.

There are instances, however, when human beings have decided that in spite of the math, we have to build new systems for society simply because living without them is just too horrible, too *emotionally* unacceptable, to consider.

In October 2002, a father and physician named Scott Gulbransen did what I do each evening when I come home—he backed into his driveway to ensure he'd have the best possible view of neighborhood children when he pulled out into the street the next morning. But he didn't see his son Cameron behind the car as he pulled in, and killed him. I'm writing this in the most clinical language I can, because frankly, it touches my worst fears as a parent, and I can't bear to go into any more detail than that. Evidently Gulbransen's story touched lawmakers' fears as well, because in 2007 Congress passed the Cameron Gulbransen Kids Transportation Safety Act, which obligated federal highway officials to figure out a rule for the design of cars that would better protect kids like Cameron.

The reasoning may have had to do with the growing realization that anyone can make the same horrific mistake Scott Gulbransen did. In a 2014 test by the Insurance Institute for Highway Safety, 111 volunteers were given the keys to a new Chevy Equinox and told to play with the infotainment system while parking the car. They were then told to back out and drive to their own car. Before they did, unbeknownst to them, a foam dummy the height of a small child was placed behind the rear bumper as if a toddler had wandered into the driveway. Without realizing it until it was too late, every single one of those 111 volunteers backed over the dummy.

It took more than a decade and a lawsuit by consumer advocate groups to finally get a rule in place, but as of 2018, any new vehicle sold in the United States weighing less than 10,000 pounds must, by law, have a backup camera installed. We recognized a terrible thing, determined that humans are inherently bad at avoiding it, and we decided not to just blame it on individuals. We recognized a society-wide problem. And perhaps most importantly, we decided that the data did *not* win the argument. If anything, the math actually *obscured* the problem.

Because here's the thing: the number of deaths prevented by backup cameras is comparatively small. A 2010 study by the National Highway Traffic Safety Administration found that backover accidents were at that point causing 292 deaths and 18,000 injuries each year on average, and that the new technology would prevent an estimated 95 of those deaths and 7,000 of those injuries. In 2018, the year backup cameras became mandatory, more than 17 million cars and light trucks were sold in the United States. Each of them carried a camera, and each of their buyers paid a little extra, in order to prevent fewer than 100 deaths. As a society we decided that 100 children being run over by their own parents is unacceptable, and that we should take steps to reduce that number, small as it is. Our emotions—our feelings as parents, the value we place on young life, the realization that *any of us* might run down our kids—won the argument.

On the other side of the coin, emotions sometimes override the data in other ways. The same year that backup cameras became mandatory, nearly 40,000 people lost their lives to a firearm in the United States, the majority—61 percent—by suicide. There was a record 323 mass shootings, including the Parkland, Florida, shooting at a high school that killed 17 students. No substantial federal legislation that year or since has changed the way the average firearm is constructed, or how their sale is regulated.

But sometimes we do it right. Who would not be moved by the horror that Cameron's parents experienced, or the simplicity of the problem they asked Congress to solve? Scott Gulbransen later wrote, "Cameron had died a sudden and horrible death because he was too small for me to see him behind my vehicle." What I'm trying to articulate here is that statistics—the data that AI would use to decide what's important—aren't always a reflection of who we are or want to be. We shouldn't always go to the top of the volcano because a guide is ready to take us there. We shouldn't plow ahead and deploy profitable technology on one another that we know has a widespread effect when we don't yet know what it is. We are capable as a society of making choices and rules based on something softer, messier, *more important* than the data, something we can all sense but can't quite see, something that touches on the human emotions we all share, the immeasurable but unmistakable dark matter of our minds.

IF WE ARE going to maintain our capacity to make these human choices, we have to incorporate the unquantifiable but unmistakable effects of The Loop into some sort of regulatory framework. It's the only way to enjoy the benefits of AI without losing human agency in the process.

We must update our legal framework to account for the gray areas between data and emotion—the heuristics and biases that AI will increasingly exploit—and come to understand that just because the effects we're feeling don't show up in a typical financial audit, they are still there, and still hugely impactful.

There was once a time in American society when companies essentially made up their own rules about the necessary precautions they should take, when the secondary effects of their businesses were considered out of their control and their internal regulations above investigation. But when enough damage was done, the law eventually got involved.

NEW YORK HARBOR, where the West Village hits the Hudson River, was once a chaos of barges. (The main thoroughfare that ran along it, 10th Avenue, was known as "Death Avenue" for the New York Central Railroad cars that would roar back and forth across pedestrian crossings, and in spite of the men hired to wave red flags from horseback to warn of oncoming trains, hundreds of New Yorkers had died under their wheels by the early 1920s.) And on a cold January morning in 1944, the tug *Carroll* was working to retrieve one of a row of barges moored to Pier 52. Barge work has always been dangerous, but at that time it carried the extra risk of being highly improvisational. When the *Carroll* approached Pier 52, her master found the one open parking space blocked by a line connecting the pier to a row of six barges, and he sent a harbormaster and a deckhand to clear the way.

The two men checked the fastenings in the cold, and then threw off the line blocking the *Carroll*'s way. Suddenly, all six barges, each more than 190 feet long, were no longer connected to the pier and drifted south. The *Carroll* and another tug tried to rescue them, but one barge, the *Anna C*, was pierced by the propeller of a nearby tanker and sank with a full load of flour owned by the US government. At that time, three months before D-Day, the war effort couldn't spare 1,500 tons of flour, much less a barge. So the government sued, and three years later, in January 1947, everyone involved was in front of the Second Circuit Court of Appeals, awaiting a ruling by Justice Learned Hand.

Justice Hand had struggled, he said, to determine whether the owners of the various barges, including the *Anna C*, should be responsible for paying someone to remain on board, so they could take charge if the barge broke away. He considered a case from a century prior in which an unmanned schooner broke loose and sank another ship. He looked at the exoneration of a powerboat owner for having no watchman aboard when a group of boys untied it for kicks in the waters off Brooklyn and it struck another vessel. He was trying to figure out whether a company that inadvertently unleashes something into shared waters should be required to keep better control of it ahead of time. "There is no general rule to determine when the absence of a bargee or other attendant will make the owner of the barge liable for injuries to other vessels if she breaks away from her moorings," he wrote. And so he came up with his own. Today, it's known as the Hand Rule.

He described it this way.

> Since there are occasions when every vessel will break from her moorings, and since, if she does, she becomes a menace to those about her; the owner's duty, as in other similar situations, to provide against resulting injuries is a function of three variables: (1) The probability that she will break away; (2) the gravity of the resulting injury, if she does; (3) the burden of adequate precautions. Possibly it serves to bring this notion into relief to state it in algebraic terms: if the probability be called P; the injury, L; and the burden, B; liability depends upon whether B is less than L multiplied by P: i. e., whether $B > PL$.[3]

Justice Hand had come up with a mathematical expression of a squishy question of human values. If I build a thing, do I have to take *every possible precaution* against the damage it might do? No, Justice Hand was saying. You only have to take precautions that cost you *less* than the harm another person might suffer without them.

In *United States v. Carroll Towing*, Justice Hand found that the cost to the company that owned the *Anna C* of paying someone to be aboard it during daylight hours was less than the cost and risk to everyone else of the barge being unattended. Today, US Coast Guard regulations require that companies have a vessel security officer for a barge, that they be on board more or less always, that they have responsibility for a Vessel Security Plan, and be in contact with the facility security officer and the vessel's company security officer, et cetera. No one is guessing which line is tied to what, and New York Harbor moved $8.4 trillion worth of goods last year.

Legal scholars have debated the use of the Hand Rule as a cost-benefit liability test for decades. Judge Richard Posner built an entire theory of the "calculus of negligence" on it, while scholars like Allan Feldman and Jeonghyun Kim have demonstrated that the rule can produce effects opposite of what Justice Hand presumably intended. Others, like Peter Grossman, have argued that the Hand Rule depends on information to which many courts simply wouldn't have access, and as a result, "parties in many cases would have no way of cost-effectively determining, *ex ante* [based on forecasts rather than results], the proper level of care to avoid liability/harm."[4]

Still, for those of us trying to sort out the responsibility of decision-technology companies to take precautions against dangerous manipulation, the amplification of primitive instincts, the loss of human agency, the Hand Rule is a useful place to start.

For one thing, it speaks not just to the intentional manipulation of human behavior, but to the unexpected secondary effects of well-intentioned businesses. If a company builds a product on pattern-recognition software that then turns out to have broken loose and gone far deeper into our lives than was intended, there will have to be a way to look back at the precautions that company should have taken. Our choices aren't just some disposable, never-ending commodity. Human agency is finite, and fragile. It costs us, in financial and emotional terms, when we lose it.

Already, courts are slowly beginning to understand that technology that shapes our decisions is having a real impact on our lives, and that in the equation B > PL, the probability and gravity of loss for all of us may far outweigh the cost of *almost any* burden on a technology company to take necessary precautions.

I asked Kathleen Wilkinson's attorney, Jay Edelson, who is suing the social casino company DoubleDown as part of a class action, whether it's hard to prove to a judge and jury that there's real harm being done to people like Wilkinson and the thousands of others in the grip of digital addiction, and he sighs.

One of his first efforts to make this sort of argument was in a Maryland District Court in 2015. He'd filed suit on behalf of another woman who'd spent more than she could afford on a game made by a Palo Alto–based company called, and I'm not kidding, Machine Zone. (That's the phrase addiction expert Natasha Dow Schüll used to describe the trance produced by compulsive slot-machine use.) Edelson lost, and in dismissing the case the court offered this opinion:

> On the surface, Plaintiff charges that Defendant trampled real and important rights and interests of hers, wrongfully and unlawfully, in an alternative, virtual world created by an electronic game. But a careful probe beneath the surface reveals a hodgepodge of hollow claims lacking allegations of real-world harms or injuries. Perceived

> unfairness in the operation and outcome of a game, where there are no real-world losses, harms, or injuries, does not and cannot give rise to the award of a private monetary remedy by a real-world court.[5]

"A lot of courts just didn't get it," Edelson says. Sure, people were spending more than they could afford. But in a free market, that's not considered a loss. That's just financial carelessness. That's the customer's problem, not society's. Telling stories like Wilkinson's, however, has helped Edelson make the case that something new is at work here, a pattern-recognition system that seeks out people whose demographics and online behavior suggest they'll be particularly susceptible to addictive play, and, once they're on the hook, does everything it can to keep them there. He points to the tactic, used by several social casino companies, of assigning "VIP hosts" to manage the relationship with the player and keep them coming back. "You have individual clients who've lost hundreds of thousands of dollars and have said to these casinos, 'I've got an addiction problem, please stop.' And then they have a specific concierge who sends them flowers on their birthday, who say, 'What can we do to get you back in the game?' Those are the stories that we're going to tell before the jury. And we're very confident it's going to resonate. Once we get these cases to the jury, we think that the casinos are in a lot of trouble."

True to his point, Edelson has begun to win. In a case against Big Fish Casino, his clients received a settlement of $155 million. With Wilkinson and her fellow plaintiffs, he says he believes he'll win again. And he says he has a new suit in the works against Apple, Google, and Facebook for essentially serving likely addicts up to social casino makers.

Why? Because those platforms are how game makers find their best customers. The most successful social casino game makers advertise on Facebook and offer a product that's playable on any smartphone. Eleven of the twelve top-grossing games on Facebook, according to Facebook's own data, are casino games.[6] Wilkinson first encountered DoubleDown Casino, Facebook's third-highest-grossing game, when the platform served her an ad for it in 2016.

According to its 2020 pre-IPO F-1 form, in 2019 DoubleDown LLC, headquartered in Korea, attracted nearly 2.9 million players per month, driving more than a quarter-billion dollars in revenue and net income of more than $36 million that year. DoubleDown's filing goes on to describe its system for getting the most money possible from its players:

> We employ a rigorous, data-driven approach to player lifecycle management from user acquisition to ongoing engagement and monetization. We use internally-developed analytic tools to segment and target players and to optimize user acquisition spend across multiple channels. Once a player is acquired, our proprietary analytic tools dissect their playing behavior on a granular level. We build a sophisticated understanding of our players that allows us to tailor game mechanics, features, and offers to drive increased gameplay or number of playing sessions. Greater engagement from our players has a flywheel effect of further improving our data analytics and ability to retain those players through customized game elements. As the players progress in their lifecycle, we also use our data analytics capabilities to boost monetization. We strategically deploy personalized special offers and tune gameplay to drive additional player spend.[7]

And Facebook openly boasts of its ability to help social casino companies like DoubleDown find the people who will be most susceptible to spending money on the game. A case study on Facebook's promotional site for likely advertising clients touts the platform's having delivered 14 percent more app downloads, at a four times lower cost per customer, for a company called PartyCasino. At a 2014 gaming conference, Julien Codorniou, who directed business for Europe, the Middle East, and Africa at Facebook at the time, and is now a Facebook vice president, told an interviewer that the company's lookalike-audience marketing product, which finds other Facebook users with interests just like a sample of your existing customers, is "very good for the gaming companies, because they can target on Facebook or on mobile specific users *or just the whales* or just the customers they want to reactivate . . . this is our business model." (Emphasis mine.) What Codorniou revealed is that Facebook was intimately familiar with the fact that some people will spend a lot of money, far more than the average person. Those are the whales, and the platforms help gaming companies find them. Facebook earns advertising revenue from those companies, as does Google. According to its policies, Apple takes a 30 percent cut of any money DoubleDown, Big Fish, or any other social casino company earning more than $1 million a year makes from in-app purchases.

All these platforms have strict rules around real-world gambling, or betting real money on digital games, as their spokespeople pointed out to me. But the law doesn't treat social casino apps as gambling, and neither do the companies' policies. (None of them would consent to an on-the-record interview about the issue, and DoubleDown Casino didn't respond to multiple requests. Big Fish wrote in a statement that its games are offered for free, for fun, and are not gambling.) Yet this is a system that is harming people, and the platforms are crucial to all of this. Without the data available to a game-maker from a platform like Facebook or Apple or Google, The Loop doesn't have what it needs to get going. With it, the cycle of behavior, analysis, winnowed choices, and compulsion can rev faster and faster.

This vast industry, comprising the game makers and the platforms that feed them customers, is built to mine our unconscious impulses for money, and the depth and sophistication of that industry means we as individuals can't hope to fight off its influence by ourselves. "I hear it a lot from legislators and the public saying 'oh, did these people just not take their statistics, their probability class in high school? Oh, we need to educate them about how chance works,'" machine-gambling expert Natasha Dow Schüll says. "This is not about people who really hope and expect to win. This is about people who discover that this game-playing is an incredibly efficient vehicle for shifting their internal moods. And you become very addicted to that, not the prospect of winning." Any of us, she argues, can fall into it.

So what should we do? Schüll, like Edelson, says we need to stop measuring "real-world" losses and start measuring the effects on our decisions. "The real value that should be measured here is more of the time spent and maybe the purchases made to continue, rather than anything about gambling or cashing out," she says. "As far as legal categories go, I think we don't have the right category for assessing the kind of harm that's done by Candy Crush, by slot machines, and by Big Fish."

How would we do that? How would we possibly put a monetary value on the squishy world of human choice? As the judge wrote in the decision in Machine Zone's favor, how do we make this more than "a hodgepodge of hollow claims"?

Happily, we don't have to build a system to analyze the value of our decisions. Facebook, among many others, has already done it.

In the early days of the Internet, advertising was sold the same way it was on television and in magazines. The given unit of measure was cost per thousand, or CPM. It denoted the number of people a television show or the publisher of *Sports Illustrated* could guarantee would see the ad, and the money you paid for an ad guaranteed you nothing more than the chance to put it in front of a fixed number of eyeballs.

Facebook, however, has invented an entirely new metric, "cost per action," and an entirely new industry, called performance marketing.

Rather than guaranteeing only that your ad will be seen by a certain number of people, Facebook will instead ask what you want to accomplish with your ad. Do you want people to buy a brassiere? Subscribe to a newsletter? Join a group? You'll pay a rate based on how many people go *do* the thing you want them to. They don't charge by the number of times the ad is seen. They charge by the actions your customer actually performs. Facebook is so proud of its ability to deliver outcomes to its advertising customers, rather than just eyeballs, that by 2018, a senior executive there told me, the majority of its advertising contracts were written on that basis.

This is a function of scale, of course. When two billion people are spending at least an hour a day on average on your platform, you can learn enough about them to deliver the ones that would want to buy a cheese knife or subscribe to a newsletter about Bitcoin.

But it's also a function of pattern recognition, and of putting a price on those patterns. If we wanted to learn the value of a potentially compulsive gambler to a social casino gaming company, we'd only need to know how much Facebook charges social casino companies each time a customer downloads an app.

In the 2015 decision against Edelson's client, the judge wrote that even if he decided to award damages to her, he'd be in the "unenviable position of pricing the conversion from virtual gold and chips to virtual wood and rock." He went on to write that "such whimsical undertaking may spark the imaginations of children and ardent game enthusiasts, but it can have no place in federal court." But the truth is that Machine Zone, like social casino companies Playtika and DoubleDown and Big Fish, knows exactly what our time is worth. They price it out every moment of the day. If we're looking for numbers to plug into the Hand Rule—i.e., the probability and gravity of a player losing more money than she can afford to spend—they're right there.

THERE IS A final human vulnerability we will have to reckon with if we're to take proper precautions against The Loop, and that's our inability to evaluate the prospect of future risks and rewards—especially when they take place far in the future.

Enrico Fermi, part of the team that created the nuclear age, finished his wartime work at Los Alamos National Lab in 1945, but would return there for a few weeks each year to kick projects around. For a time he worked on the question of what happens at the border between two types of liquid, for instance. But by 1950 he was back at work on weapons, designing the hydrogen bomb after the Soviet Union detonated a fission weapon the year prior.

That summer he fell into a regular lunchtime quartet, four scientists just nerding out together on escapist topics with meal trays in front of them. At one of these, the group was discussing the sheer number of planets there must be across the universe, and several of them remember Fermi asking some version of a fundamental question: "Where is everyone?"

And with that, he launched an entire academic field of dreamy astrophysics. Fermi's Paradox, as it has come to be known, is the idea that there are so many stars in the sky, representing so much potentially life-enabling energy, that life must be out there somewhere. There are an estimated 200 billion galaxies in the universe, leading to conservative estimates that the number of planets could top 700 quintillion or more. In 1961, the astrophysicist Frank Drake estimated that between 1,000 and 100,000,000 planets in the Milky Way alone probably have some sort of civilization on them. Drake went on to found SETI, the planet's premiere nongovernmental effort to look for alien life. As a 2019 University of Rochester paper explained it, "If technologically advanced exo-civilizations are common, then we should already have evidence of their existence either through direct or indirect means."[8] So where *is* everyone?

In 1973, the MIT astronomer John Allen Ball proposed that perhaps alien civilizations are holding back on purpose, waiting for humanity to become, well, more fun to be around. Carl Sagan wondered whether alien thought may be too slow or perhaps too fast for us to detect. Some papers have suggested that every civilization must go through a "great filter" like climate change or a pandemic that destroys or spares them, and that we're the only ones in the neighborhood to have survived ours. But my favorite explanation is the one first put forth in 1961 by the German astronomer Sebastian von Hoerner. He arrived, by some napkin math, at the idea that the average civilization would likely last no more than 6,500 years, and that such civilizations were likely 1,000 light-years apart. Von Hoerner was writing at the height of the Cold War, when the threat of nuclear annihilation was on everyone's minds, and his theory had more to do with what he considered the likely self-destructive nature of any civilization than with the time and distance between one and the next. But his idea helped launch a new way of looking at the paradox.

Perhaps the question isn't "Where is everyone?" It's "When is everyone?" Because in the vastness of space and time, we are the briefest flare of a matchhead against the darkness. Our species in any form has only been alive for a tiny sliver of our planet's 4.5-billion-year existence. And who is to say how many matches have been struck before us, and will be struck again once we're gone? Species on other worlds may have crawled ashore and developed civilizations and died off into dust billions of years before we did the same, and billions of years after our planet is rid of us it may happen again elsewhere. It would be an outrageous coincidence to think that in the hugeness of the universe two matches might flare at the same moment. As the astrobiologist Caleb Scharf wrote in *Scientific American* in 2014, "On the truly grand scale of things the existence of a self-aware, technological species could always be a fleeting event, a wisp reaching out from the tumult of molecular evolution, to be inevitably reabsorbed or redirected."[9]

We're built to assume that the most significant events and effects are the ones that take place in front of and upon us, when in fact it may be the ones that happened long ago or will happen eons from now that shape our species. This is why threats like pandemics and climate change are so terribly dangerous to humanity. If we do not personally *experience* rising waves or the sensation of struggling to breathe from our beds in the grip of a virus, we can't access our most efficient and effective mental systems for avoiding danger. But by the time the waves hit our houses or we're being put on a ventilator, it's too late to do anything. We are not good with abstract concepts across large amounts of time, and that inability to see beyond our own life spans is what makes The Loop possible. When we can't form conversations with our coworkers without AI's suggestions, when music is only composed from the patterns in what we've chosen before, when whole industries are endlessly sampling the same ever-shrinking pool of human choice, it will be too late. This is a species-wide problem. It will affect all of us. We have to put some guidelines in place now.

With some focusing constraints and a clear understanding of what it can and can't do, we might use The Loop for good, rather than just turning it loose on the population for any profitable pattern it can find. Rather than amplifying our most unthinking qualities, and shrinking our choices in the process, perhaps we can use decision technology to compensate for our shortcomings, and get The Loop spinning such that it expands our choices.

We've seen how this can work with something like home insurance, and there are other examples to inspire optimism. Perhaps it takes the form of a system that forces us to be acquainted with the issues on our ballots, and forces the people seeking our vote to care about them, too. Pol.is, a simple platform for analyzing the patterns in online opinions, has been adapted by the Taiwanese government to better inform its elected officials as to what their constituents really want. Pol.is offers a list of propositions and asks respondents to rank their feelings about each one. It then analyzes clusters of sentiment and delivers the equivalent of a white paper to government officials as part of a program called vTaiwan. For a couple of years, it was even the backbone of an online politics show, *Talk to Taiwan*, in which host Chang Tieh-chih brutally grilled guests like Taipei mayor Ko Wen-je using highly specific questions generated by data. Colin Megill told me he created Pol.is because he was tired of people showing up at town halls, exhausting themselves for hours learning a subject, and then never returning. As he told me, "You'd never have had something like Brexit if people had used something like this, because it would have been clear, right there, what voters actually wanted. There'd be no way for parliament to misrepresent public opinion."

Perhaps AI should only be applied to public problems at first. Green River AI is one of the very few firms that restricts its deployment of machine learning to public health and social justice. But since he founded the company in 2000, Michael Knapp told me, he's discovered that without the inherent accelerating efficiencies of greed, AI isn't being deployed where it's needed. "My frustration is that I can figure out a child is likely to be lead poisoned from their birth certificate, and I can develop a system for going into the home three years ahead of time and get things fixed. But the social-services system won't let me do that *until* that kid is already damaged for the rest of their lives." When I asked Knapp whether he worried about decision technology and AI collapsing human choice, he laughed, and told me he wished he had that problem. In his world, AI presents a vast, untapped opportunity for positive change, but nonprofits and city governments and social-service agencies just don't have the resources to take advantage of the possibilities, even while companies like Facebook run whole academic-style labs devoted to it. "That's why we're so zealous about this technology," Knapp said. "It's getting used effectively on the dark side. It's not being used well on this other side."

But the larger question of this moment is whether we're going to let the institutions that will most immediately profit turn AI loose on us, or whether there's something gritty and difficult and important in our decision-making apparatus that we want to protect and perhaps even improve. Our individual and social patterns are already built to give away our agency to mechanisms outside our consciousness—to our assumptions, to cues in our surroundings, to our peers—whenever we possibly can. We hoard mental energy even when we don't need to, even when we *shouldn't.* As so many researchers into System 1 and System 2 and bias and emotion-led decision-making have found, we are literally built to avoid thinking for ourselves. This is what has set The Loop in motion, and we're on the precipice of giving our entire way of life over to an invisible, imperceptible, irresistible instinct to hand difficult choices—even and especially the vitally important ones—to automated systems. We know, from decades of science, that we have the power to improve ourselves by making conscious choices about our lives and by looking at our vulnerabilities clearly. We know, from centuries of struggle, that we can build institutions that help us rise above the worst parts of our ancient selves. Now we have to learn, in this moment, how to resist the pull of convenience and profit and protect the best aspects of who we are.

ACKNOWLEDGMENTS

SOMEDAY IN THE future, if the world I see coming is actually coming, a writer will essentially dictate a few thoughts into a piccc of software, and it will go on to write a complete book for her. And the convenience won't stop there. That author will also have the option of allowing the software, or some all-encompassing platform on which she does everything anyway, to look across her communications and her address book and automatically assemble a heartfelt note of gratitude to all the people it detects have contributed to her thinking. I find myself wishing at this moment that such a system existed to assemble for me the long, complete list of all the people who helped get this book done, because I'm terrified of forgetting someone. If you're that person, please accept my gratitude anyway.

First, I'd like to thank the hundreds of people over the years who have spoken to me off the record about the places they work, especially those at multinational corporations. The landscape that journalists and sources at major companies have to traverse to find one another has on the one hand become logistically easier, in that I can reach someone in any country in the world in moments. But the standard information-suppression strategy at your average technology company involves threatening the careers of those who speak to the press without permission. And that, combined with a strategy of actively indoctrinating employees about why their work is on the right side of history (which they somehow square with forbidding employees from speaking about that work with outsiders), makes people who will share with me what they know very hard to find and very important. I believe the future is being determined inside companies that do not want to reveal their work until it's too late for the public to have an effect on it. People at those companies who risk so much to speak with the press help to rebalance that very lopsided dynamic. I'm enormously grateful to those who have entrusted me with that responsibility.

This book was made possible by the many friends and colleagues who have listened to me worry about human nature and technology all these years, and have helped me shape all of my high hopes and deep anxieties into something approaching a cogent theory. I took a long walk in a cemetery with the producer and director Carl Byker in 2015, and began a friendship that took me all over the world and taught me all about the ancient guidance systems we unconsciously obey in our lives. His insight, patience, and unrelenting good cheer are a model to me and got me through this often painful writing process. I'm also very grateful to producer Kate McMahon, who introduced me to some of the most influential behavioral scientists in the world.

The Center for Advanced Study in the Behavioral Sciences at Stanford University, along with the Berggruen Institute, funded a year of writing this book, and I'm indebted to its director, the political scientist Margaret Levi, for taking a gamble on me, and to the other fellows for their consideration and generosity with their time and expertise. The writer and entrepreneur Tim O'Reilly first suggested I apply to CASBS, and his generous enthusiasm for my subject helped me realize there might be a book in it.

I continue to be astonished at the good fortune I've had to work at NBC News, where from my first interview I've found creative collaborators to help me pursue my strange, dark subject on television. Noah Oppenheim, Janelle Rodriguez, Rashida Jones, Libby Leist, Tom Mazzarelli, Elena Nachmanoff, Jessica Kurdali, Dan Arnall, Betsy Korona, Polly Powell, Audrey Grady, Chiara Sottile, and my colleagues in the Business, Technology, and Media Unit and at bureaus around the world have made it possible for me to pursue stories like the ones in these pages, although on the face of it these sorts of stories offer almost nothing at which to point a camera. I'm deeply grateful to Heather Allan, who gave me my first opportunity in television news and taught me from the first day to treat the technology beat as being about the inequities of capitalism rather than about the shiny doodads of the future. David Doss showed me how to do the job. And Evan Groll helped me build the habits of mind and body I needed to be able to put something like this together.

Popular Science's editor in chief Mark Jannot gave me the job that set me on my professional path, and eventually he handed his own job to me, a priceless gift. I'd also like to thank the many friends and colleagues I found in the magazine industry, who fought to preserve and improve long-form journalism even as the business model, gutted by social media, collapsed around them.

I'm grateful to my television agent Peter Goldberg and literary agent Byrd Leavell for their trust and good counsel, and to my editor, Sam Raim, for shaping what began as a disjointed series of essays into something resembling a book. The book's mistakes are my responsibility. Its strengths are his.

No one understands that work like this is even possible until they've seen someone they admire doing it first, and I'm very lucky to have a family that has set that example for me. My father, the author Andrew Ward, carved out a very futuristic career writing about the injustices of his own people decades before it was fashionable to do so, and raised my sister and me at home while doing it. Thanks, Dad. My mother, the public health academic Deborah Ward, taught me about doing the needful, as her mother called it, and helped me understand one of this book's major themes, that good decisions usually feel wrong, even when they're right. Nathan Ward taught me to find pleasure and dignity in journalism. Garrett Ward taught me not to be embarrassed by the things I find interesting. Geoffrey Ward taught me to look for big subjects and great collaborators.

My children, Josephine and Juniper, somehow found it in themselves to forgive me, at least outwardly, as I repeatedly abandoned them after dinner, on weekends, and for long stretches of travel for the years it took me to get this done. My most painful regret in writing this book is the time it took away from them.

The luck I often speak of in describing my time as a journalist—the opportunity, the privilege, the satisfaction—is all code for the things my wife Julie makes possible. There is nothing I've accomplished that doesn't directly depend on her encouragement, love, and sacrifice. I hope that I can someday support her dreams the way she has supported mine.

NOTES

CHAPTER 1: THE REALITY GAP

1. Ghazal, Y.A., Stahnisch, F.W. (2015). Otto Poetzl (1877–1962). *Journal of Neurology*, 262(3):795–797. doi: 10.1007/s00415-015-7679-6

2. Erdelyi, M. (1984). The recovery of unconscious (inaccessible) memories: Laboratory studies of hypermnesia. *Psychology of Learning and Motivation*, Vol. 18, 95–127. doi: 10.1016/s0079-7421(08)60360-6

3. Weindling, P. (2021). *From Clinic to Concentration Camp: Reassessing Nazi Medical and Racial Research, 1933–1945*. New York and London: Routledge.

4. Frankl, V.E. (2000). *Recollections: An Autobiography*. Cambridge: Basic Books.

5. Goodale, M., Milner, D. (November 2006). One brain—two visual systems. *The Psychologist*, Vol. 19, 660–663, https://thepsychologist.bps.org.uk/volume-19/edition-11/one-brain-two-visual-systems

6. Eagleman, D. (2009). Brain time. In M. Brockman, ed., *What's Next? Dispatches on the Future of Science*. New York: Vintage Books.

7. de Gelder, B., Vroomen, J., Pourtois, G., Weiskrantz, L. (1999). Non-conscious recognition of affect in the absence of striate cortex. *Neuroreport*, Dec 16; 10(18):3759–3763. doi: 10.1097/00001756-199912160-00007. PMID: 10716205.

CHAPTER 2: ILLUSIONS

1. Fischhoff, B. (1975). Hindsight ≠ foresight: The effect of outcome knowledge on judgment under uncertainty. *Journal of Experimental Psychology: Human Perception and Performance*, Vol. 1, 288–299.

2. James, W. (2001). *Talks to Teachers on Psychology and to Students on Some of Life's Ideals*. New York: Dover Publications.

3. Slovic, P., Peters, E. (2006). Risk perception and affect. *Current Directions in Psychological Science*, 15(6):322–325. doi: 10.1111/j.1467-8721.2006.00461.x

CHAPTER 3: TWO SYSTEMS

1. Kahneman, D. (2003). A perspective on judgment and choice: Mapping bounded rationality. *American Psychologist*, 58(9):697–720. doi: 10.1037/0003-066X.58.9.697

2. Ambady, N. (2010). The perils of pondering: Intuition and thin slice judgments. *Psychological Inquiry*, 21(4):271–278. doi: 10.1080/1047840X.2010.524882

3. Todorov, A., Mandisodza, A.N., Goren, A., Hall, C.C. (2005). Inferences of competence from faces predict election outcomes. *Science*, Jun 10; 308(5728):1623–1626. doi: 10.1126/science.1110589. PMID: 15947187.

CHAPTER 4: CLUSTERS

1. Terrizzi, J.A., Jr., Shook, N.J., McDaniel, M.A. (2013). The behavioral immune system and social conservatism: A meta-analysis. *Evo-*

lution and Human Behavior, 34(2):99–108, https://linkinghub.elsevier.com/retrieve/pii/S109051381200116X

2. Young, J.L., Hegarty, P. (2019). Reasonable men: Sexual harassment and norms of conduct in social psychology. *Feminism & Psychology*, 29(4):453–474. doi: 10.1177/0959353519855746

3. Gómez, Á., López-Rodríguez, L., Sheikh, H., et al. (2017). The devoted actor's will to fight and the spiritual dimension of human conflict. *Nature Human Behavior*, 1:673–679. doi: 10.1038/s41562-017-0193-3

4. Hamid, N., Pretus, C., Atran, S., Crockett, M.J., et al. (2019). Neuroimaging 'will to fight' for sacred values: An empirical case study with supporters of an Al Qaeda associate. *Royal Society Open Science*, Jun 12; 6(6):181585. doi: 10.1098/rsos.181585

CHAPTER 5: GUIDANCE SYSTEMS

1. Mezulis, A.H., Abramson, L.Y., Hyde, J.S., Hankin, B.L. (2004). Is there a universal positivity bias in attributions? A meta-analytic review of individual, developmental, and cultural differences in the self-serving attributional bias. *Psychological Bulletin*, 130(5):711–747.

2. Waldenström, U., Schytt, E. (2009). A longitudinal study of women's memory of labour pain—from 2 months to 5 years after the birth. *BJOG*, Mar; 116(4):577–583. doi: 10.1111/j.1471-0528.2008.02020.x. Epub 2008 Dec 9. PMID: 19076128.

3. Gainsbury, S.M., Russell, A.M.T., King, D.L., Delfabbro, P., Hing, N. (2016). Migration from social casino games to gambling: Motivations and characteristics of gamers who gamble. *Computers in Human Behavior*, 63:59–67. doi: 10.1016/j.chb.2016.05.021

4. The 192nd Central Court of the Commonwealth of Massachusetts. Session Laws, Acts (2011), Chapter 194. "An Act Establishing Expanded Gaming in the Commonwealth." https://malegislature.gov/Laws/SessionLaws/Acts/2011/Chapter194

5. Carden, L., Wood, W. (2018). Habit formation and change. *Current Opinion in Behavioral Sciences*, 20:117–122. doi: 10.1016/j.cobeha.2017.12.009

CHAPTER 6: LIFE ON THE RAILS

1. Erickson, P., et al. (2015). *How Reason Almost Lost Its Mind: The Strange Career of Cold War Rationality.* Chicago: University of Chicago Press.

2. Waytz, A., Morewedge, C.K., Epley, N., Monteleone, G., Gao, J.H., Cacioppo, J.T. (2010). Making sense by making sentient: Effectance motivation increases anthropomorphism. *Journal of Personality and Social Psychology*, 99(3):410–435. doi: 10.1037/a0020240

3. Risen, J.L. (2016). Believing what we do not believe: Acquiescence to superstitious beliefs and other powerful intuitions. *Psychological Review*, 123(2):182–207. doi: 10.1037/rev0000017

4. Weizenbaum, J. (1976). *Computer Power and Human Reason: From Judgment to Calculation*. San Francisco: W. H. Freeman and Company.

5. Colby, K.M., Watt, J.B., Gilbert, J.P. (1966). A computer method of psychotherapy: Preliminary communication. *Journal of Nervous and Mental Disease*, 142(2):148–152.

6. Sagan, C. (January 1975). Cosmic calendar. *Natural History*, 84(10):70–73.

7. ben-Aaron, D. (1985, April 9). Interview: Weizenbaum examines computers and society. *The Tech*, http://tech.mit.edu/V105/N16/weisen.16n.html/

8. http://dig.abclocal.go.com/wls/documents/2018/041018-wls-iteam-united-complaint-doc.pdf

9. "The priority of all other confirmed passengers may be determined based on a passenger's fare class, itinerary, status of frequent flyer program membership, whether the passenger purchased the ticket under select UA corporate travel agreements, and the time in which the passenger presents him/herself for check-in without advanced seat assignment." www.united.com/ual/en/us/fly/contract-of-carriage.html

CHAPTER 7: WHAT AI ISN'T

1. http://raysolomonoff.com/dartmouth/notebook/dart56ipl_01 17.pdf

2. Rudin, C., Radin, J. (2019). Why are we using black box models in AI when we don't need to? A lesson from an explainable AI competition. *Harvard Data Science Review*, 1(2). doi: 10.1162/99608f92 .5a8a3a3d

3. Fulton, R., Holsinger, B., eds. (2007). *History in the Comic Mode: Medieval Communities and the Matter of Person*. New York: Columbia University Press.

4. As summarized by Whittaker, Sir Edmund (1989). *A History of the Theories of Aether and Electricity: Vol. I: The Classical Theories; Vol. II: The Modern Theories, 1900–1926*. Mineola, NY: Dover Publications.

5. O'Hara, J.G., Pricha, W. (1987). *Hertz and the Maxwellians: A Study and Documentation of the Discovery of Electromagnetic Wave Radiation, 1873–1894*. London: P. Peregrinus Limited.

6. Fölsing, A. (1997). *Albert Einstein: A Biography*. New York: Viking.

7. Kronovet, D. (2017, March 28). Objective functions in machine learning. *Abacus* (blog), http://kronosapiens.github.io/blog/2017/03/28 /objective-functions-in-machine-learning.html

8. Patty, J.W., Penn, E.M. (2014). *Social Choice and Legitimacy: The Possibilities of Impossibility*. New York: Cambridge University Press. doi: 10.1017/CBO9781139030885

9. Nou, J. (2015). Review of social choice and legitimacy. *The New Rambler*, https://newramblerreview.com/images/files/Jennifer_Nou -Review_Patty-and-Penn.pdf

CHAPTER 9: THE LOOP

1. Mangel, M., Samaniego, F.J. (1984). Abraham Wald's work on aircraft survivability. *Journal of the American Statistical Association*, 79(386):259–267. doi: 10.1080/01621459.1984.10478038

2. Tenbarge, K. (2021, May 16). Sorry Bella Poarch, this IS "Build a B*tch." *The Kids Aren't Alright* (newsletter on Substack), https://kidsarentalright.substack.com/p/sorry-bella-poarch-this-is-build

3. Kelly, S.D. (2019, February 21). A philosopher argues that an AI can't be an artist. *MIT Technology Review.*

4. Brundage, M. [@Miles_Brundage], (2020, December 28). "It's hard to say which classes or specific use cases will spread the most. But it seems safe to say that the era of almost-exclusively-human-generated and almost-never-individually-customized media will not last much longer, absent consumer/policy demand for it continuing. /Fin," [Tweet] Twitter. https://twitter.com/Miles_Brundage/status/1343750861047009288?s=20

5. Buolamwini, J., Gebru, T. (2018). Gender shades: Intersectional accuracy disparities in commercial gender classification. *Proceedings of the 1st Conference on Fairness, Accountability and Transparency*, PMLR 81:77–91.

6. Perdomo, J.C., Zrnic, T., Mendler-Dünner, C., Hardt, M. (February 2020). Performative prediction. *ArXiv*, https://arxiv.org/abs/2002.06673

7. Baude, A., Pearson, J., Drapeau, S. (2016). Child adjustment in joint physical custody versus sole custody: A meta-analytic review. *Journal of Divorce Remarriage*, 57(5):338–360. doi: 10.1080/10502556.2016.1185203

8. Bauserman, R. (2012). A meta-analysis of parental satisfaction, adjustment, and conflict in joint custody and sole custody following divorce. *Journal of Divorce Remarriage*, 53:464–488. doi: 10.1080/10502556.2012.682901

9. Hirsch, J. (2019, September 2). Your boss is going to start using AI to monitor you—and labor laws aren't ready. *Fast Company*, www.fastcompany.com/90397285/worker-protection-laws-arent-ready-for-the-boom-in-corporate-surveillance-of-employees

10. Franaszek, K. (2021). Tens of thousands of news articles are labeled as unsafe for advertisers. *Adalytics*, adalytics.io/blog/tens-of-thousands-of-news-articles-are-labeled-as-unsafe-for-advertisers

CHAPTER 10: MISSION CRITICAL

1. Eterno, J.A., Verma, A., Silverman, E.B. (2014). Police manipulations of crime reporting: Insiders' revelations. *Justice Quarterly*, 33(5), 811–835. doi: 10.1080/07418825.2014.980838

2. Nagin, D.S. (2013). Deterrence in the 21st century: A review of the evidence. In M. Tonry, ed., *Crime and Justice: An Annual Review of Research*. Chicago: University of Chicago Press.

3. Friedman, B. (2018, June 22). The worrisome future of policing technology. *New York Times*, www.nytimes.com/2018/06/22/opinion/the-worrisome-future-of-policing-technology.html

4. Lum, K., Isaac, W. (2016). To predict and serve? *Significance*, 13:14–19. doi: 10.1111/j.1740-9713.2016.00960.x

5. Papachristos, A.V., Wildeman, C., Roberto, E. (2015). Tragic, but not random: The social contagion of nonfatal gunshot injuries. *Social Science & Medicine*, Jan; 125:139–150. doi: 10.1016/j.socscimed.2014.01.056

6. Zhao, L., Papachristos, A.V. (2020). Network position and police who shoot. *The Annals of the American Academy of Political and Social Science*, 687(1):89–112. doi: 10.1177/0002716219901171

7. Rickert, C. (2018, June 11). City insurance costs rise with police shooting payouts. *Wisconsin State Journal*, https://madison.com/wsj/news/local/crime/city-insurance-costs-rise-with-police-shooting-payouts/article_322439f5-32ff-5aaf-a44a-00f47f12931a.html

CHAPTER 11: WEAK PERFECTION

1. Cuéllar, M.-F. (2016, October 1). Cyberdelegation and the Administrative State. Stanford Public Law Working Paper No. 2754385.

2. California Penal Code Chapter 4. Plea. Section 1016–1027.

3. Gilman, M. (2020, September 15). Poverty Lawgorithms: A Poverty Lawyer's Guide to Fighting Automated Decision-Making Harms on Low-Income Communities. Data & Society Research Institute, https://datasociety.net/library/poverty-lawgorithms

4. Arnold, D., Dobbie, W., Yang, C.S. (May 2017). Racial bias in bail decisions. NBER Working Paper No. 23421, Cambridge: National Bureau of Economic Research, www.nber.org/system/files/working_papers/w23421/w23421.pdf

5. DeMichele, M., Baumgartner, P., Wenger, M., Barrick, K., Comfort, M. (2020). Public safety assessment: Predictive utility and differential prediction by race in Kentucky. *Criminology & Public Policy*, 19:409–431. doi: 10.1111/1745-9133.12481

6. Pretrial Justice Institute, Updated Position on Pretrial Risk Assessment Tools (2020, February 7), www.pretrial.org/wp-content/uploads/Risk-Statement-PJI-2020.pdf

7. https://twitter.com/CA_EDD/status/1345766205257224192?s=20

8. Greenberg, D. (1983). Donald Black's sociology of law: A critique. *Law & Society Review*, 17:337. doi: 10.2307/3053351

9. United States Sentencing Commission. (2017). Demographic Differences in Sentencing: An Update to the 2012 Booker Report, www.ussc.gov/sites/default/files/pdf/research-and-publications/research-publications/2017/20171114_Demographics.pdf

CHAPTER 12: HIGHER MATH

1. Verisk. (2020). Wildfire risk analysis, www.verisk.com/insurance/campaigns/location-fireline-state-risk-report

2. Hirschman, A.O. (2007). How the exit option works. *Exit, Voice, and Loyalty: Responses to Decline in Firms, Organizations, and States*. Cambridge: Harvard University Press.

3. United States v. Carroll Towing Co., 159 F.2d 169 (2d Cir. 1947).

4. Grossman, P.Z., Cearley, R.W., Cole, D.H. (March 2006). Uncertainty, insurance and the Learned Hand formula. *Law, Probability and Risk*, 5(1):1–18. doi: 10.1093/lpr/mgl012

5. http://cdn.arstechnica.net/wp-content/uploads/2015/10/gameofwar.pdf

6. www.facebook.com/games/query/top-grossing/

7. DoubleDown Interactive Co., Ltd. (2020). 2019 F-1, p. 85, www.sec.gov/Archives/edgar/data/1799567/000119312520158606/d753829df1.htm

8. Carroll-Nellenback, J., Frank, A., Wright, J., Scharf, C. (2019). The Fermi Paradox and the Aurora Effect: Exo-civilization settlement, expansion, and steady states. *The Astronomical Journal*, 158(3):117. doi: 10.3847/1538-3881/ab31a3

9. Scharf, C.A. (2014, March 20). The unstoppable extinction and Fermi's paradox. *Scientific American*, https://blogs.scientificamerican.com/life-unbounded/the-unstoppable-extinction-and-fermie28099s-paradox/